Learn, Practice, Succeed

Eureka Math®

Grade 8
Module 3

Published by Great Minds®.

Copyright © 2019 Great Minds®.

Printed in the U.S.A.

This book may be purchased from the publisher at eureka-math.org.

10 9 8 7 6 5 4 3 2

ISBN 978-1-64054-982-1

G8-M3-LPS-05.2019

Students, families, and educators:

Thank you for being part of the *Eureka Math*® community, where we celebrate the joy, wonder, and thrill of mathematics.

In *Eureka Math* classrooms, learning is activated through rich experiences and dialogue. That new knowledge is best retained when it is reinforced with intentional practice. The *Learn, Practice, Succeed* book puts in students' hands the problem sets and fluency exercises they need to express and consolidate their classroom learning and master grade-level mathematics. Once students learn and practice, they know they can succeed.

What is in the Learn, Practice, Succeed *book?*

Fluency Practice: Our printed fluency activities utilize the format we call a Sprint. Instead of rote recall, Sprints use patterns across a sequence of problems to engage students in reasoning and to reinforce number sense while building speed and accuracy. Sprints are inherently differentiated, with problems building from simple to complex. The tempo of the Sprint provides a low-stakes adrenaline boost that increases memory and automaticity.

Classwork: A carefully sequenced set of examples, exercises, and reflection questions support students' in-class experiences and dialogue. Having classwork preprinted makes efficient use of class time and provides a written record that students can refer to later.

Exit Tickets: Students show teachers what they know through their work on the daily Exit Ticket. This check for understanding provides teachers with valuable real-time evidence of the efficacy of that day's instruction, giving critical insight into where to focus next.

Homework Helpers and Problem Sets: The daily Problem Set gives students additional and varied practice and can be used as differentiated practice or homework. A set of worked examples, Homework Helpers, support students' work on the Problem Set by illustrating the modeling and reasoning the curriculum uses to build understanding of the concepts the lesson addresses.

Homework Helpers and Problem Sets from prior grades or modules can be leveraged to build foundational skills. When coupled with *Affirm*™, *Eureka Math*'s digital assessment system, these Problem Sets enable educators to give targeted practice and to assess student progress. Alignment with the mathematical models and language used across *Eureka Math* ensures that students notice the connections and relevance to their daily instruction, whether they are working on foundational skills or getting extra practice on the current topic.

Where can I learn more about Eureka Math *resources?*

The Great Minds® team is committed to supporting students, families, and educators with an ever-growing library of resources, available at eureka-math.org. The website also offers inspiring stories of success in the *Eureka Math* community. Share your insights and accomplishments with fellow users by becoming a *Eureka Math* Champion.

Best wishes for a year filled with "aha" moments!

Jill Diniz

Jill Diniz
Chief Academic Officer, Mathematics
Great Minds

Students, families, and educators

Contents

Module 3: Similarity

©2019 Great Minds®. eureka-math.org

Exploratory Challenge

Two geometric figures are said to be similar if they have the same shape but not necessarily the same size. Using that informal definition, are the following pairs of figures similar to one another? Explain.

Pair A:

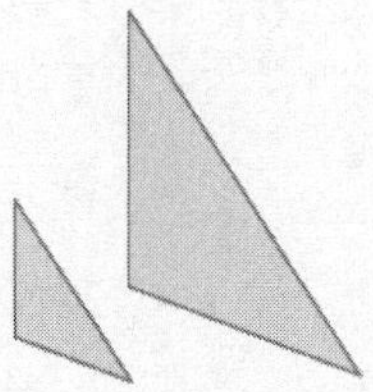

Pair B:

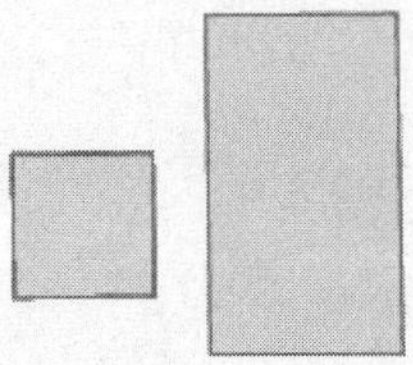

Pair C:

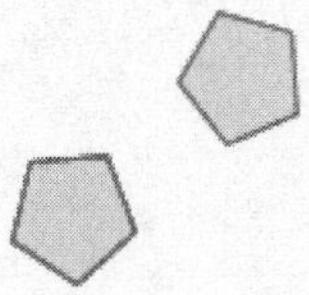

Pair D:

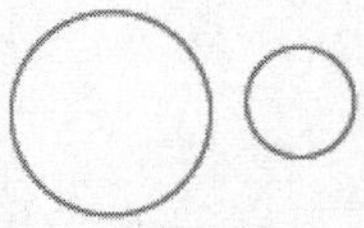

©2019 Great Minds®. eureka-math.org

Pair E:

Pair F:

Pair G:

Pair H:

Exercises

1. Given $|OP| = 5$ in.
   a. If segment OP is dilated by a scale factor $r = 4$, what is the length of segment OP'?

   b. If segment OP is dilated by a scale factor $r = \frac{1}{2}$, what is the length of segment OP'?

©2019 Great Minds®. eureka-math.org

Use the diagram below to answer Exercises 2–6. Let there be a dilation from center O. Then, $Dilation(P) = P'$ and $Dilation(Q) = Q'$. In the diagram below, $|OP| = 3 \text{ cm}$ and $|OQ| = 4 \text{ cm}$, as shown.

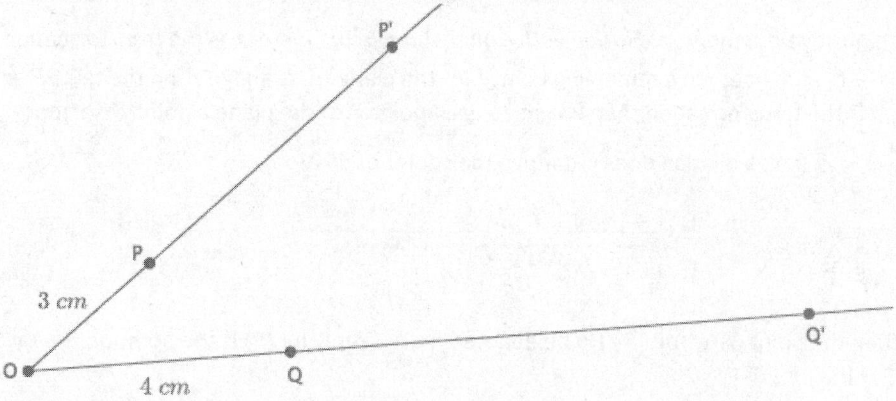

P'
P
3 cm
O
4 cm
Q
Q'

2. If the scale factor is $r = 3$, what is the length of segment OP'?

3. Use the definition of dilation to show that your answer to Exercise 2 is correct.

4. If the scale factor is $r = 3$, what is the length of segment OQ'?

5. Use the definition of dilation to show that your answer to Exercise 4 is correct.

6. If you know that $|OP| = 3$, $|OP'| = 9$, how could you use that information to determine the scale factor?

©2019 Great Minds®. eureka-math.org

Lesson Summary

Definition: For a positive number r, a *dilation with center O and scale factor r* is the transformation of the plane that maps O to itself, and maps each remaining point P of the plane to its image P'on the ray $\overrightarrow{OP}$ so that $|OP'| = r|OP|$. That is, it is the transformation that assigns to each point P of the plane a point $Dilation(P)$ so that

1. $Dilation(O) = O$ (i.e., a dilation does not move the center of dilation).

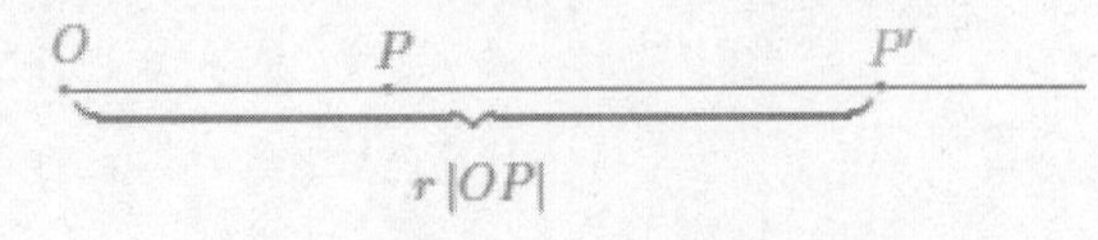

2. If $P \neq O$, then the point $Dilation(P)$ (to be denoted more simply by P') is the point on the ray $\overrightarrow{OP}$ so that $|OP'| = r|OP|$.

In other words, a dilation is a rule that moves each point P along the ray emanating from the center O to a new point P' on that ray such that the distance $|OP'|$ is r times the distance $|OP|$.

©2019 Great Minds®. eureka-math.org

Name ______________________________ Date ______________

1. Why do we need a better definition for similarity than "same shape, not the same size"?

2. Use the diagram below. Let there be a dilation from center O with scale factor $r = 3$. Then, $Dilation(P) = P'$. In the diagram below, $|OP| = 5\text{ cm}$. What is $|OP'|$? Show your work.

O
5cm
P
P'

3. Use the diagram below. Let there be a dilation from center O. Then, $Dilation(P) = P'$. In the diagram below, $|OP| = 18\text{ cm}$ and $|OP'| = 9\text{ cm}$. What is the scale factor r? Show your work.

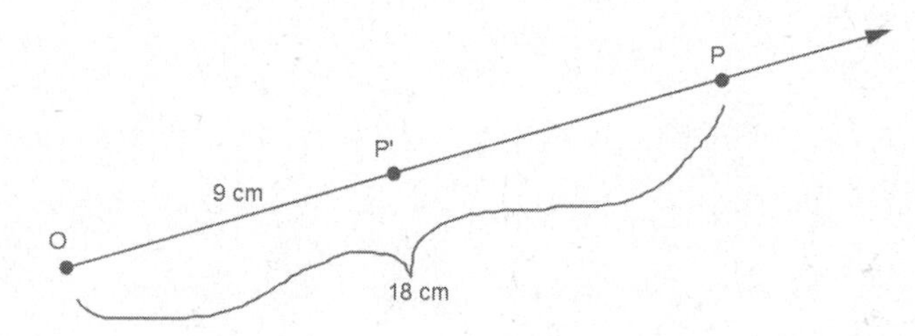

©2019 Great Minds®. eureka-math.org

1. Let there be a dilation from center O. Then, $Dilation(P) = P'$, and $Dilation(Q) = Q'$. Examine the drawing below. What can you determine about the scale factor of the dilation?

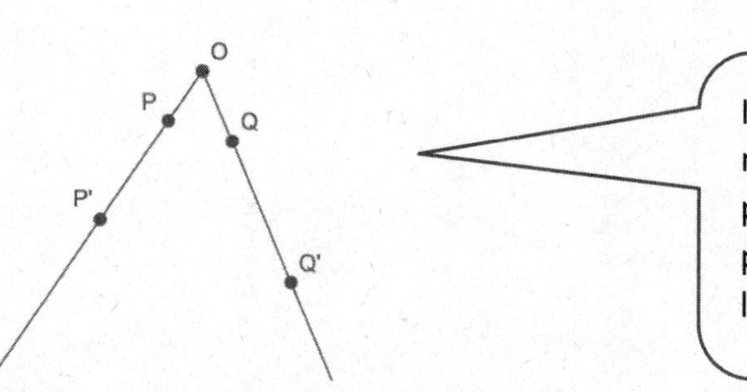

I remember from the last module that the original points are labeled without primes, and the images are labeled with primes.

The dilation must have a scale factor larger than 1, $r > 1$, since the dilated points are farther from the center than the original points.

2. Let there be a dilation from center O with a scale factor $r = 2$. Then, $Dilation(P) = P'$, and $Dilation(Q) = Q'$. $|OP| = 1.7$ cm, and $|OQ| = 3.4$ cm, as shown. Use the drawing below to answer parts (a) and (b). The drawing is not to scale.

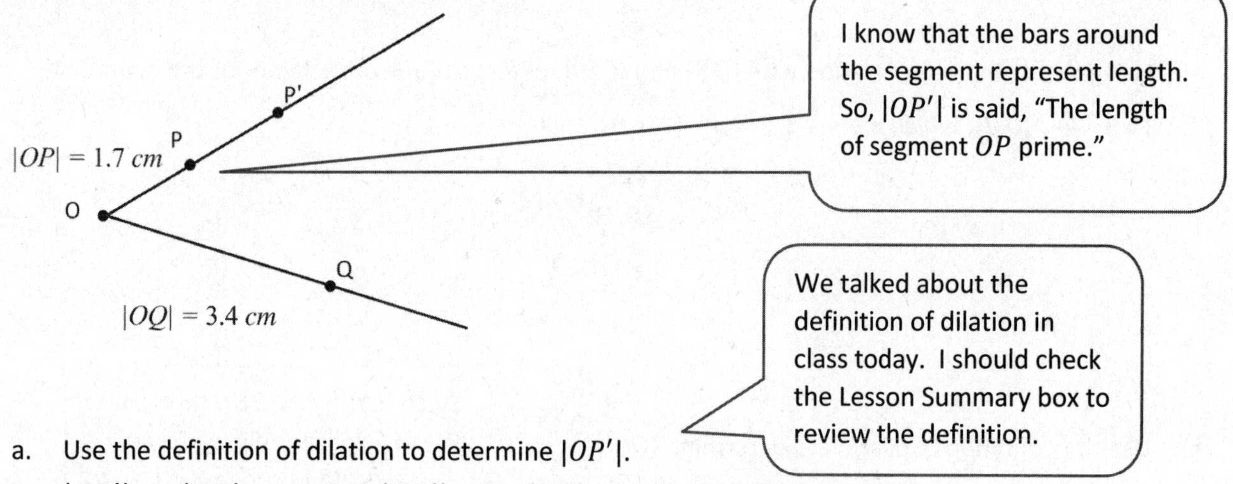

   a. Use the definition of dilation to determine $|OP'|$.
   $|OP'| = r|OP|$; therefore, $|OP'| = 2 \cdot (1.7) = 3.4$ and $|OP'| = 3.4$ cm.

   b. Use the definition of dilation to determine $|OQ'|$.
   $|OQ'| = r|OQ|$; therefore, $|OQ'| = 2 \cdot (3.4) = 6.8$ and $|OQ'| = 6.8$ cm.

©2019 Great Minds®. eureka-math.org

3. Let there be a dilation from center O with a scale factor r. Then, $Dilation(B) = B'$, $Dilation(C) = C'$, and $|OB| = 10.8$ cm, $|OC| = 5$ cm, and $|OB'| = 2.7$ cm, as shown. Use the drawing below to answer parts (a)–(c).

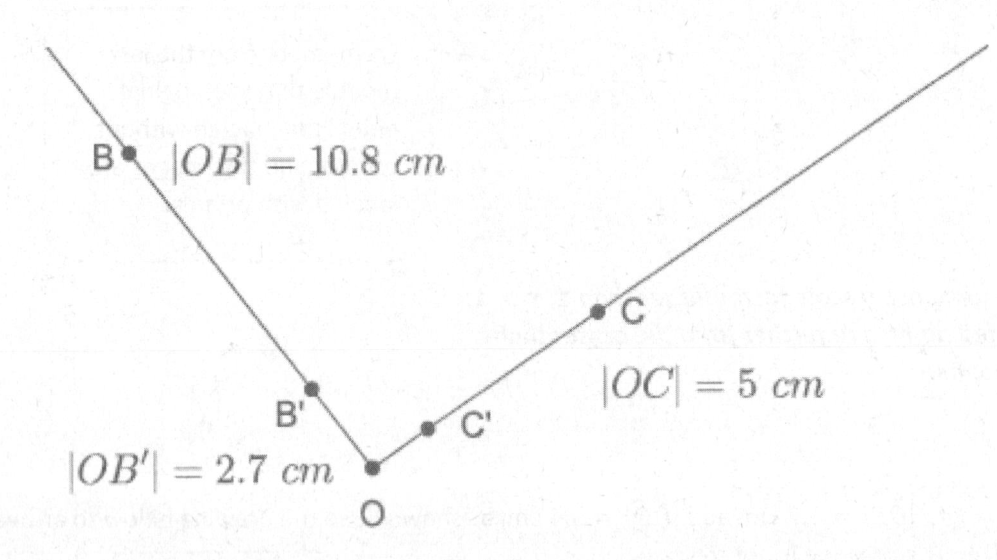

a. Using the definition of dilation with $|OB|$ and $|OB'|$, determine the scale factor of the dilation.

$|OB'| = r|OB|$, which means $2.7 = r \cdot (10.8)$, then

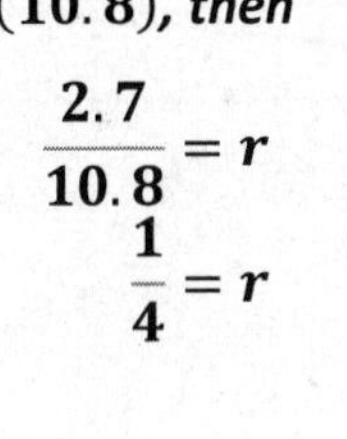

$$\frac{2.7}{10.8} = r$$
$$\frac{1}{4} = r$$

Since $|OB'| = r|OB|$, then by the multiplication property of equality, $\frac{|OB'|}{|OB|} = r$.

b. Use the definition of dilation to determine $|OC'|$.

Since the scale factor, r, is $\frac{1}{4}$, then $|OC'| = \frac{1}{4}|OC|$; therefore, $|OC'| = \frac{1}{4} \cdot 5 = 1.25$, and $|OC'| = 1.25$ cm.

©2019 Great Minds®. eureka-math.org

1. Let there be a dilation from center O. Then, $Dilation(P) = P'$ and $Dilation(Q) = Q'$. Examine the drawing below. What can you determine about the scale factor of the dilation?

O Q Q' P P'

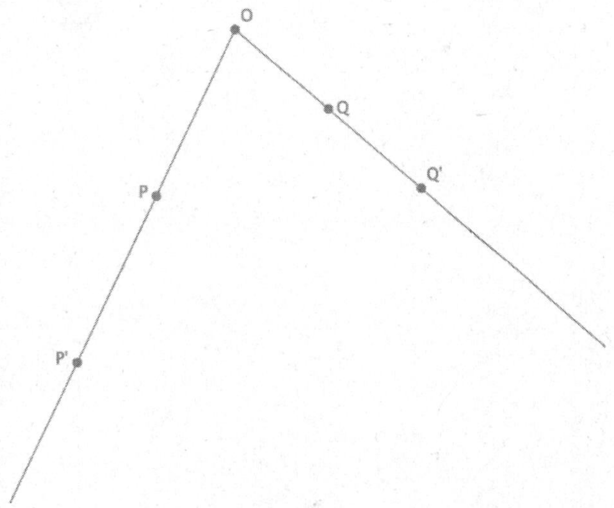

2. Let there be a dilation from center O. Then, $Dilation(P) = P'$, and $Dilation(Q) = Q'$. Examine the drawing below. What can you determine about the scale factor of the dilation?

O Q' P' Q P

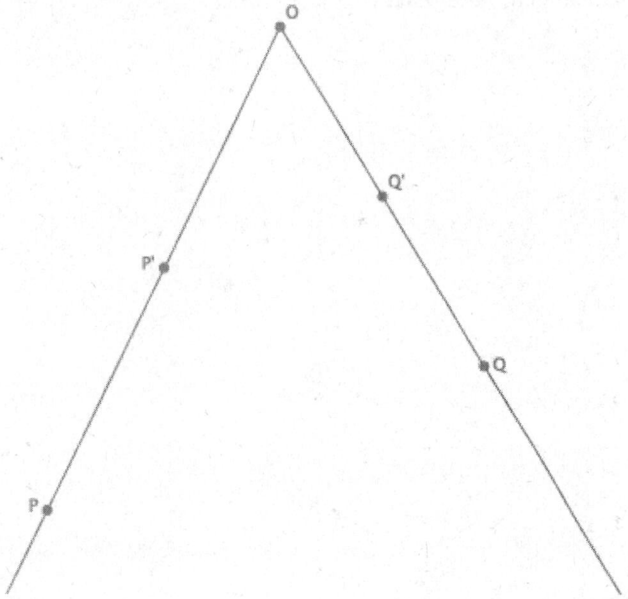

©2019 Great Minds®. eureka-math.org

3. Let there be a dilation from center O with a scale factor $r = 4$. Then, $Dilation(P) = P'$ and $Dilation(Q) = Q'$. $|OP| = 3.2$ cm, and $|OQ| = 2.7$ cm, as shown. Use the drawing below to answer parts (a) and (b). The drawing is not to scale.

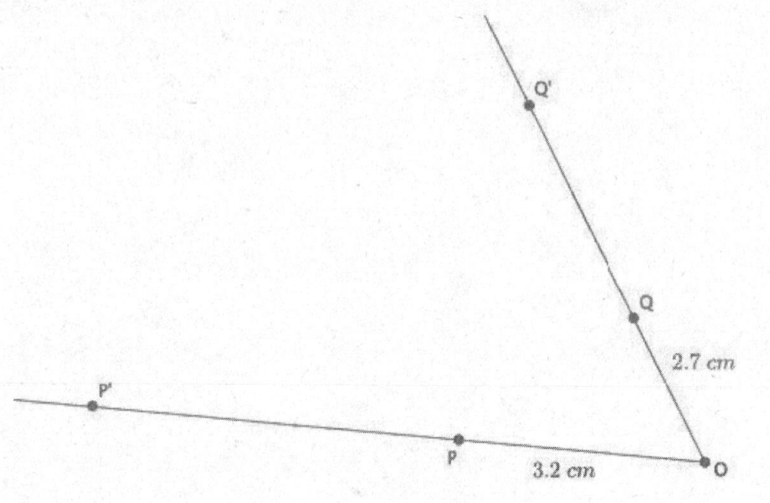

a. Use the definition of dilation to determine $|OP'|$.

b. Use the definition of dilation to determine $|OQ'|$.

©2019 Great Minds®. eureka-math.org

4. Let there be a dilation from center O with a scale factor r. Then, $Dilation(A) = A'$, $Dilation(B) = B'$, and $Dilation(C) = C'$. $|OA| = 3$, $|OB| = 15$, $|OC| = 6$, and $|OB'| = 5$, as shown. Use the drawing below to answer parts (a)–(c).

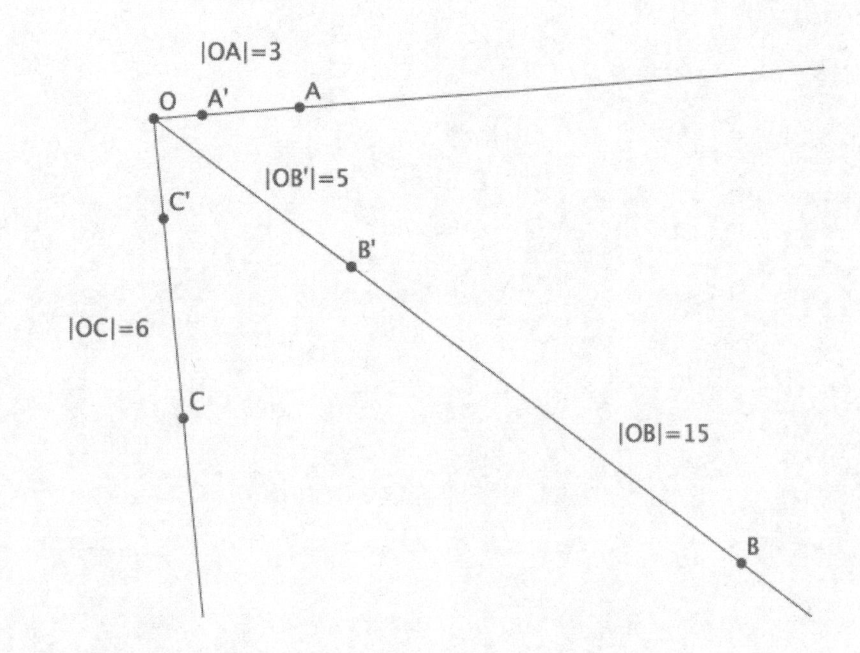

a. Using the definition of dilation with lengths OB and OB', determine the scale factor of the dilation.

b. Use the definition of dilation to determine $|OA'|$.

c. Use the definition of dilation to determine $|OC'|$.

©2019 Great Minds®. eureka-math.org

Examples 1–2: Dilations Map Lines to Lines

©2019 Great Minds®. eureka-math.org

Example 3: Dilations Map Lines to Lines

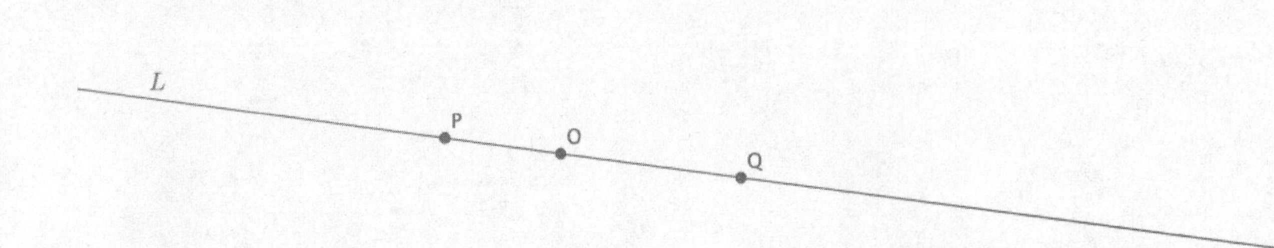

Exercise

Given center O and triangle ABC, dilate the triangle from center O with a scale factor $r = 3$.

a. Note that the triangle ABC is made up of segments AB, BC, and CA. Were the dilated images of these segments still segments?

©2019 Great Minds®. eureka-math.org

b. Measure the length of the segments AB and $A'B'$. What do you notice? (Think about the definition of dilation.)

c. Verify the claim you made in part (b) by measuring and comparing the lengths of segments BC and $B'C'$ and segments CA and $C'A'$. What does this mean in terms of the segments formed between dilated points?

d. Measure $\angle ABC$ and $\angle A'B'C'$. What do you notice?

e. Verify the claim you made in part (d) by measuring and comparing the following sets of angles: (1) $\angle BCA$ and $\angle B'C'A'$ and (2) $\angle CAB$ and $\angle C'A'B'$. What does that mean in terms of dilations with respect to angles and their degrees?

©2019 Great Minds®. eureka-math.org

Lesson Summary

Dilations map lines to lines, rays to rays, and segments to segments. Dilations map angles to angles of the same degree.

©2019 Great Minds®. eureka-math.org

Name ____ Date ____________________

1. Given center O and quadrilateral $ABCD$, using a compass and ruler, dilate the figure from center O by a scale factor of $r = 2$. Label the dilated quadrilateral $A'B'C'D'$.

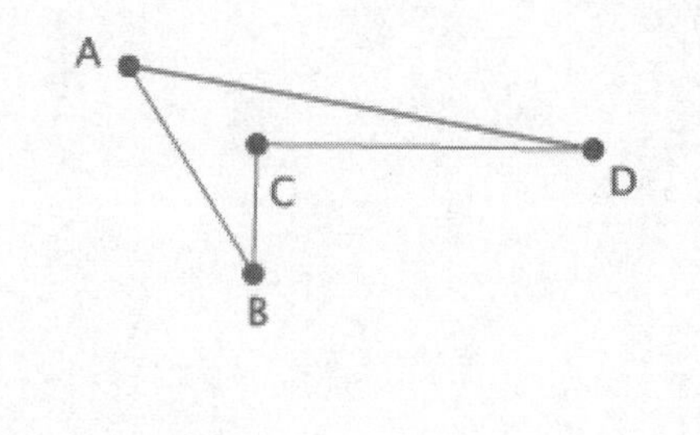

2. Describe what you learned today about what happens to lines, segments, rays, and angles after a dilation.

©2019 Great Minds®. eureka-math.org

1. Given center O and quadrilateral $ABCD$, use a ruler to dilate the figure from center O by a scale factor of $r = \frac{1}{4}$. Label the dilated quadrilateral $A'B'C'D'$.

O

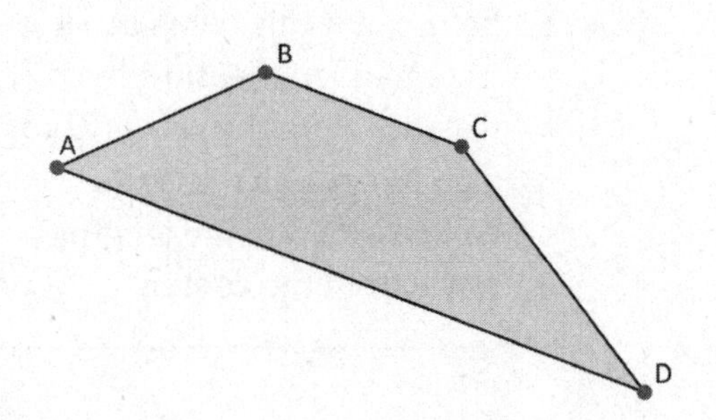

I need to draw the rays from the center O to each point on the figure and then measure the length from O to each point.

The figure in red below shows the dilated image of $ABCD$. All measurements are in centimeters.

$\lvert OA'\rvert = r\lvert OA\rvert$	$\lvert OB'\rvert = r\lvert OB\rvert$	$\lvert OC'\rvert = r\lvert OC\rvert$	$\lvert OD'\rvert = r\lvert OD\rvert$
$\lvert OA'\rvert = \frac{1}{4}(11.2)$	$\lvert OB'\rvert = \frac{1}{4}(9.2)$	$\lvert OC'\rvert = \frac{1}{4}(7.6)$	$\lvert OD'\rvert = \frac{1}{4}(6.8)$
$\lvert OA'\rvert = 2.8$	$\lvert OB'\rvert = 2.3$	$\lvert OC'\rvert = 1.9$	$\lvert OD'\rvert = 1.7$

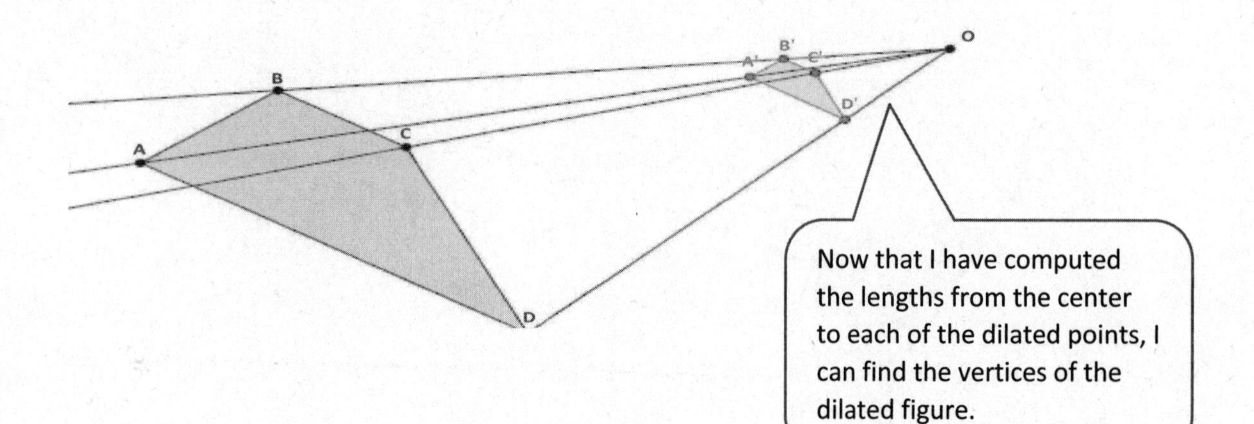

Now that I have computed the lengths from the center to each of the dilated points, I can find the vertices of the dilated figure.

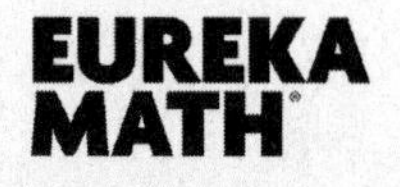

©2019 Great Minds®. eureka-math.org

2. Use a compass to dilate the figure ABC from center O, with scale factor $r = 3$.

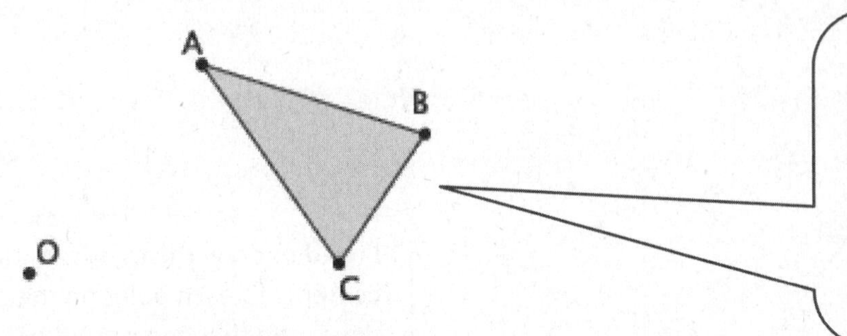

I need to draw the rays like before, but this time I can use a compass to measure the distance from the center O to a point and again use the compass to find the length 3 times from the center.

The figure in red below shows the dilated image of ABC.

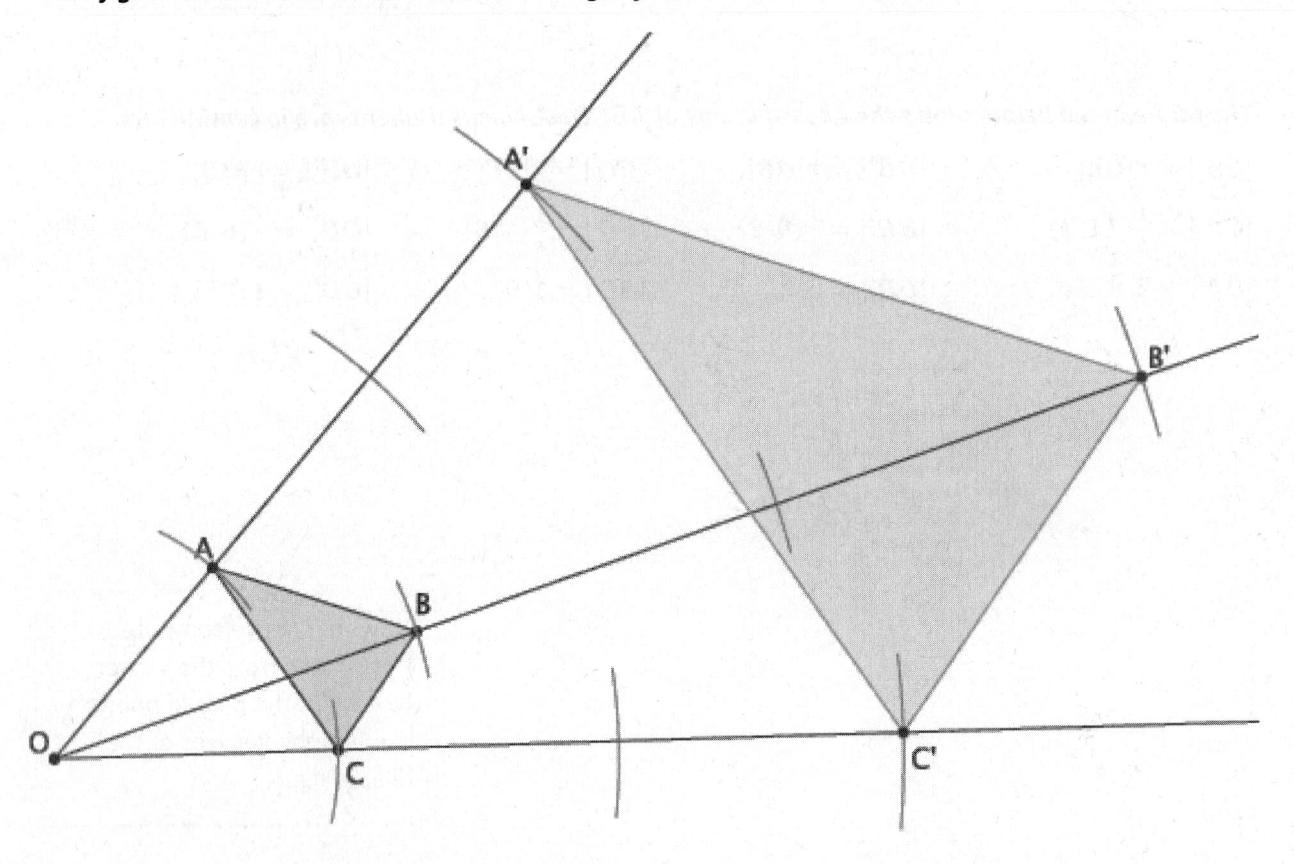

©2019 Great Minds®. eureka-math.org

3. Use a compass to dilate the figure $ABCD$ from center O, with scale factor $r = 2$.

   a. Dilate the same figure, $ABCD$, from a new center, O', with scale factor $r = 2$. Use double primes ($A''B''C''D''$) to distinguish this image from the original.

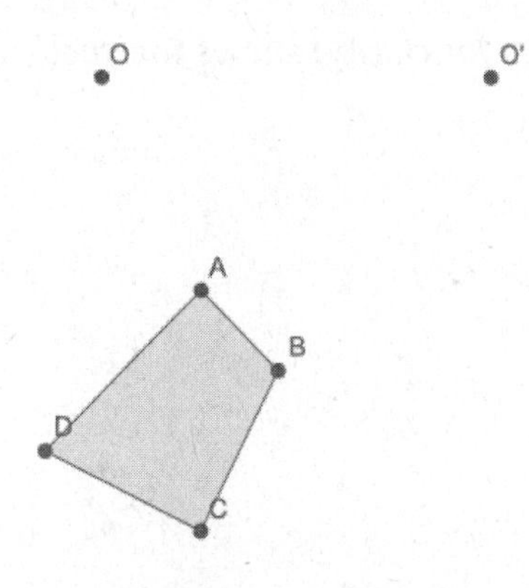

The figure in blue, $A'B'C'D'$, shows the dilation of $ABCD$ from center O, with scale factor $r = 2$.
The figure in red, $A''B''C''D''$, shows the dilation of $ABCD$ from center O' with scale factor $r = 2$.

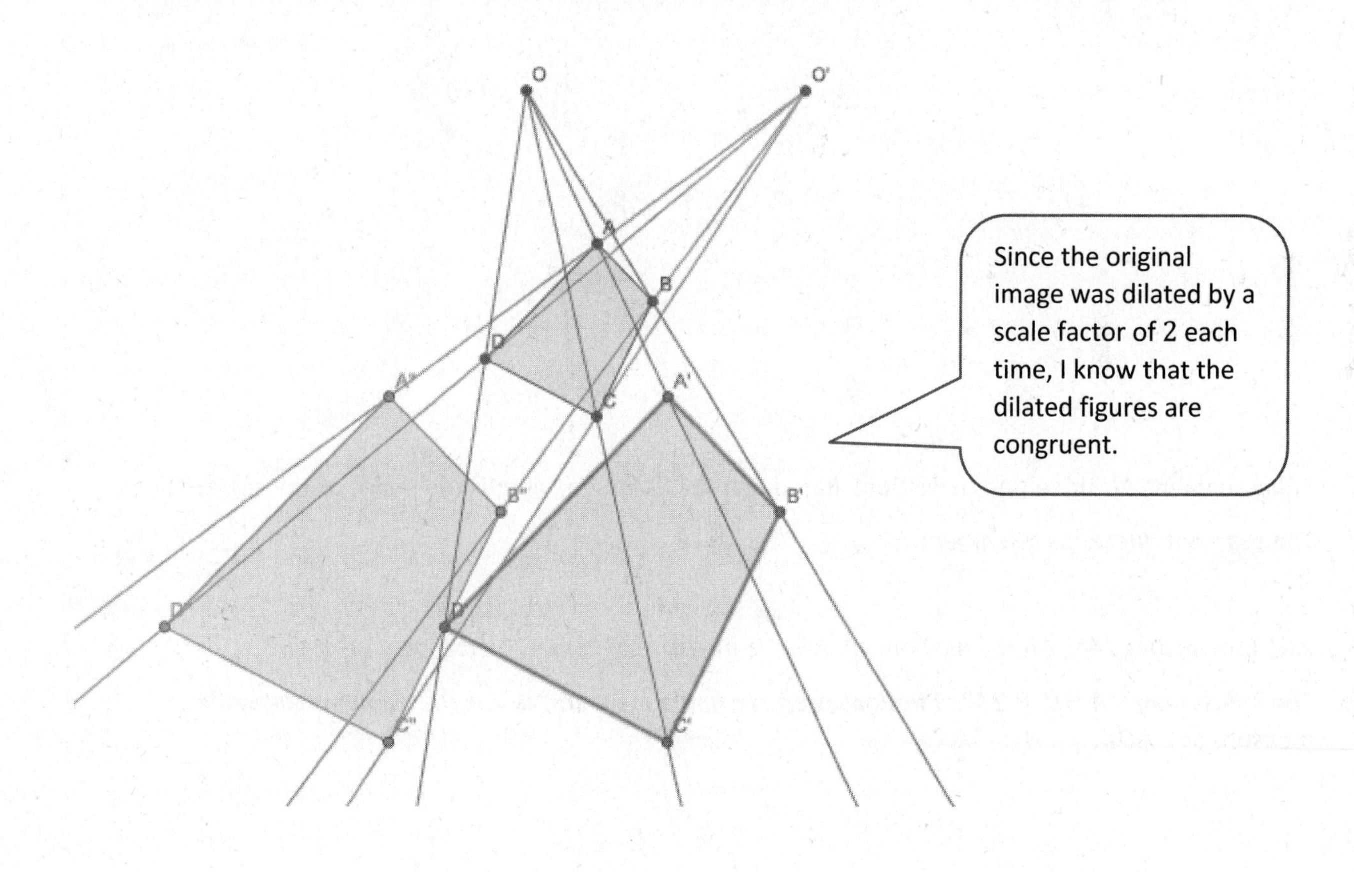

©2019 Great Minds®. eureka-math.org

b. What rigid motion, or sequence of rigid motions, would map $A''B''C''D''$ to $A'B'C'D'$?

A translation along vector $\overrightarrow{A''A'}$ (or any vector that connects a point of $A''B''C''D''$ and its corresponding point of $A'B'C'D'$) would map the figure $A''B''C''D''$ to $A'B'C'D'$.

The image below (with rays removed for clarity) shows the vector $\overrightarrow{A''A'}$.

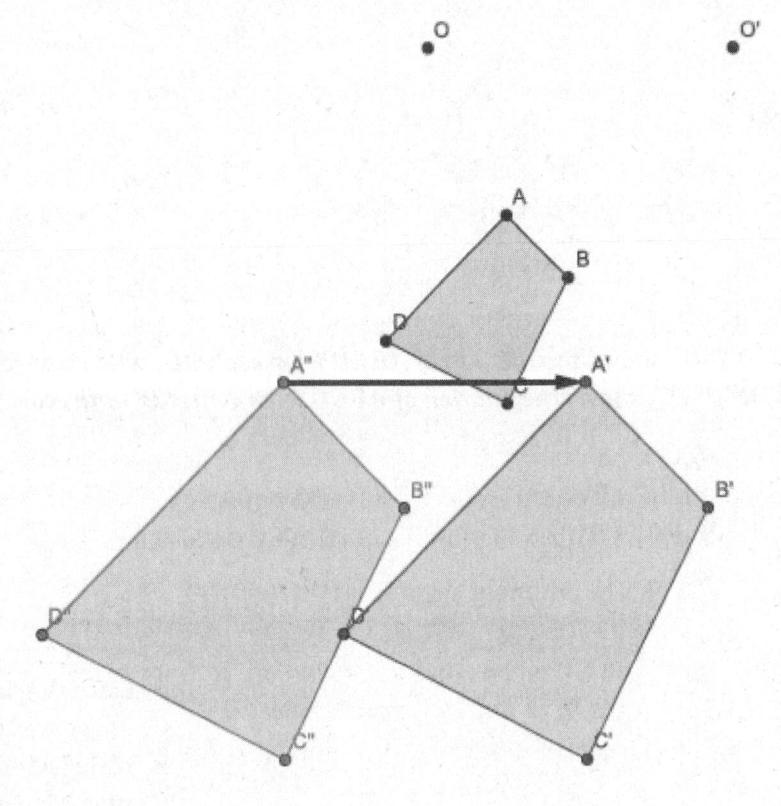

4. A line segment AB undergoes a dilation. Based on today's lesson, what is the image of the segment?

The segment dilates as a segment.

5. $\angle AOC$ measures $24°$. After a dilation, what is the measure of $\angle A'OC'$? How do you know?

The measure of $\angle A'OC'$ is $24°$. Dilations preserve angle measure, so $\angle A'OC'$ remains the same measure as $\angle AOC$.

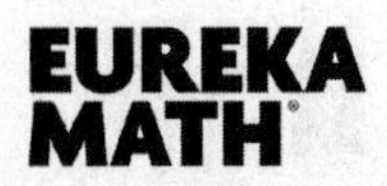

©2019 Great Minds®. eureka-math.org

1. Use a ruler to dilate the following figure from center O, with scale factor $r = \frac{1}{2}$.

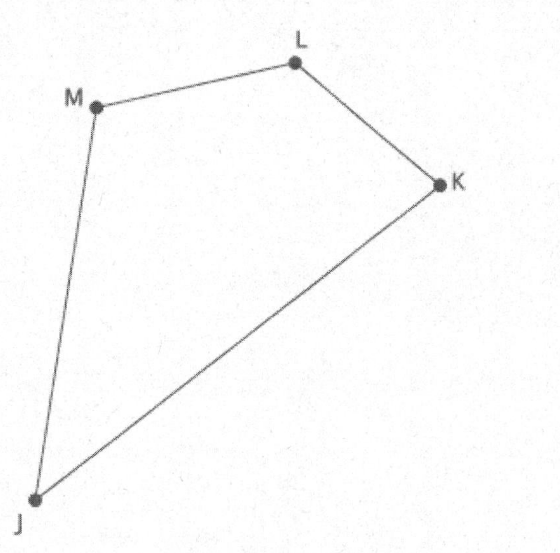

©2019 Great Minds®. eureka-math.org

2. Use a compass to dilate the figure $ABCDE$ from center O, with scale factor $r = 2$.

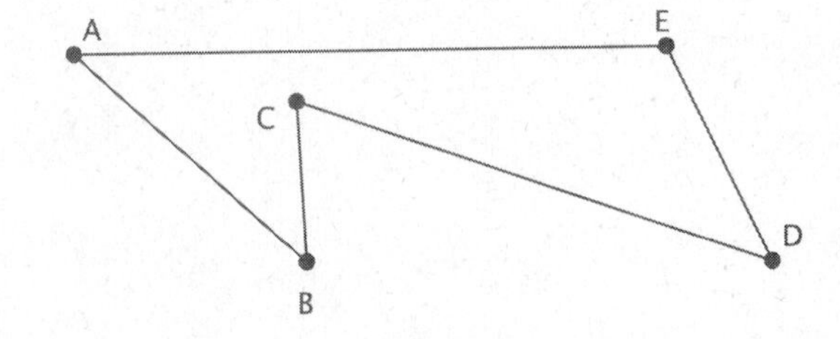

O'

O

a. Dilate the same figure, $ABCDE$, from a new center, O', with scale factor $r = 2$. Use double primes ($A''B''C''D''E''$) to distinguish this image from the original.

b. What rigid motion, or sequence of rigid motions, would map $A''B''C''D''E''$ to $A'B'C'D'E'$?

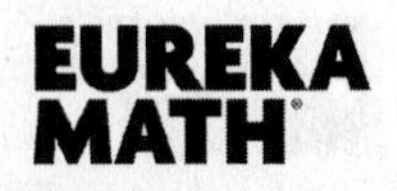

©2019 Great Minds®. eureka-math.org

3. Given center O and triangle ABC, dilate the figure from center O by a scale factor of $r = \frac{1}{4}$. Label the dilated triangle $A'B'C'$.

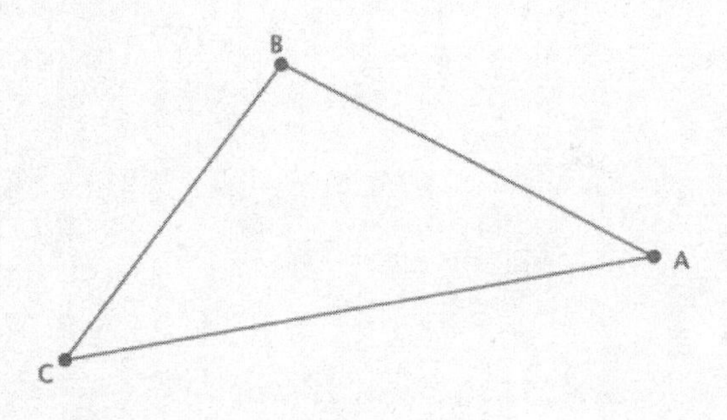

O

4. A line segment AB undergoes a dilation. Based on today's lesson, what is the image of the segment?

5. $\angle GHI$ measures $78°$. After a dilation, what is the measure of $\angle G'H'I'$? How do you know?

©2019 Great Minds®. eureka-math.org

Dilate circle A from center O at the origin by scale factor $r = 3$.

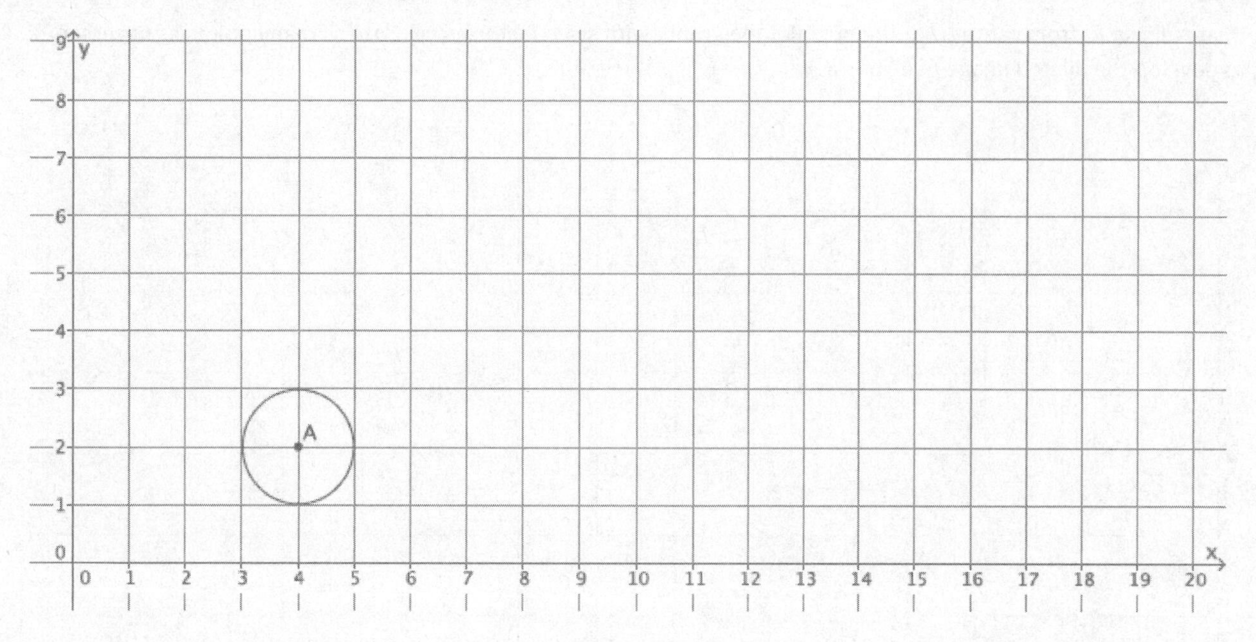

©2019 Great Minds®. eureka-math.org

Exercises 1–2

1. Dilate ellipse E, from center O at the origin of the graph, with scale factor $r = 2$. Use as many points as necessary to develop the dilated image of ellipse E.

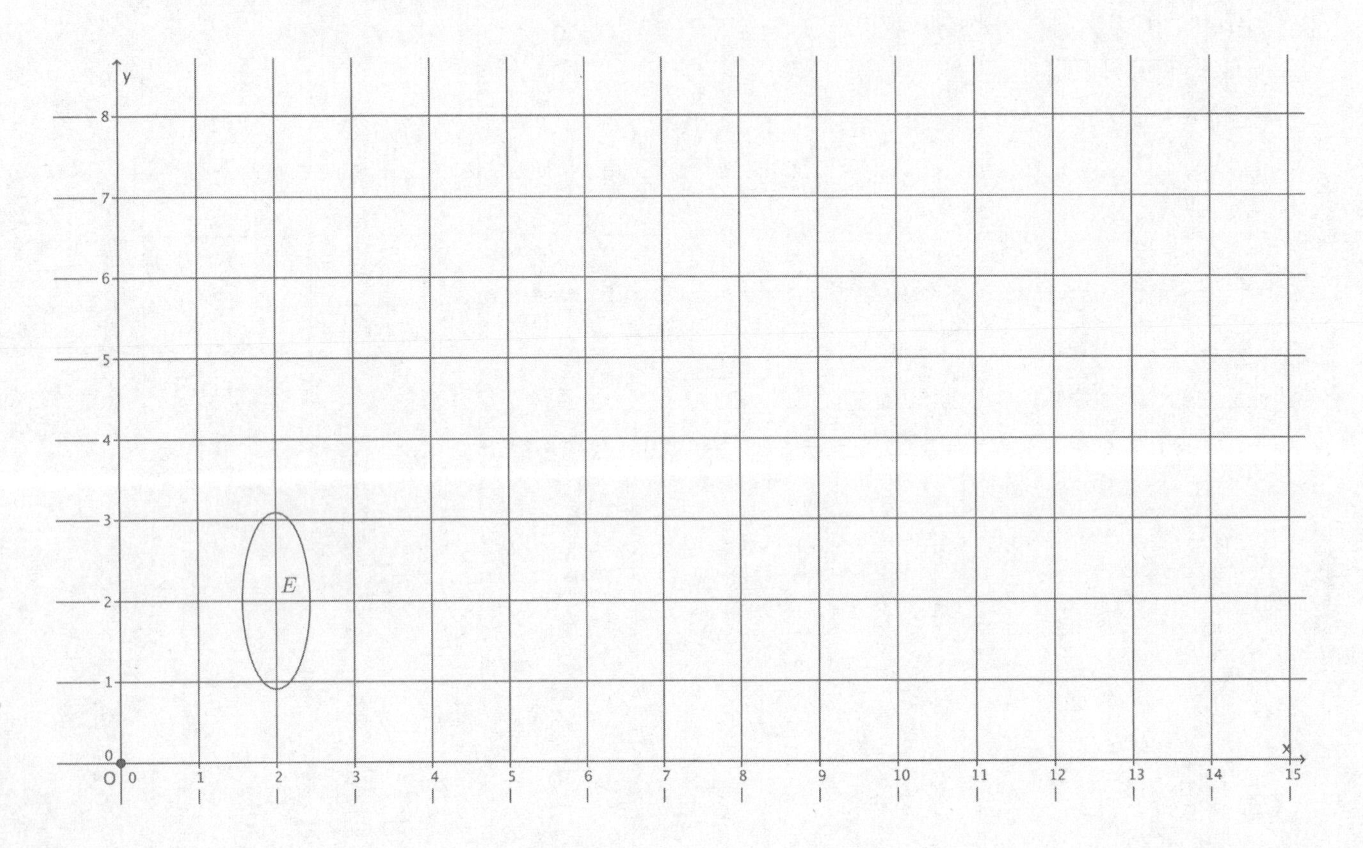

2. What shape was the dilated image?

©2019 Great Minds®. eureka-math.org

Exercise 3

3. Triangle ABC has been dilated from center O by a scale factor of $r = \frac{1}{4}$ denoted by triangle $A'B'C'$. Using a centimeter ruler, verify that it would take a scale factor of $r = 4$ from center O to map triangle $A'B'C'$ onto triangle ABC.

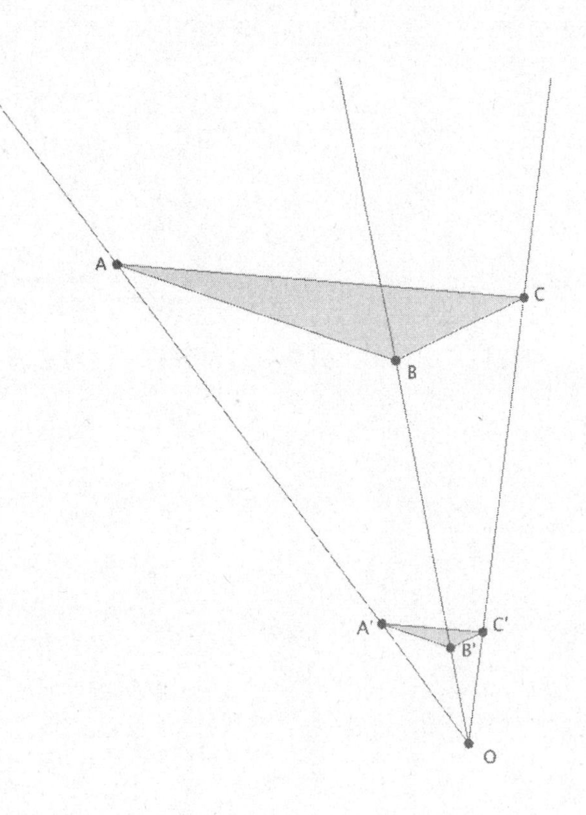

©2019 Great Minds®. eureka-math.org

Lesson Summary

Dilations map circles to circles and ellipses to ellipses.

If a figure is dilated by scale factor r, we must dilate it by a scale factor of $\frac{1}{r}$ to bring the dilated figure back to the original size. For example, if a scale factor is $r = 4$, then to bring a dilated figure back to the original size, we must dilate it by a scale factor $r = \frac{1}{4}$.

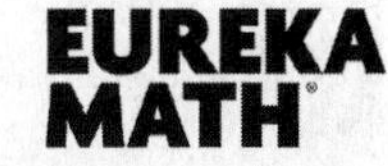

©2019 Great Minds®. eureka-math.org

Name ______________________________ Date ______________

1. Dilate circle A from center O by a scale factor $r = \frac{1}{2}$. Make sure to use enough points to make a good image of the original figure.

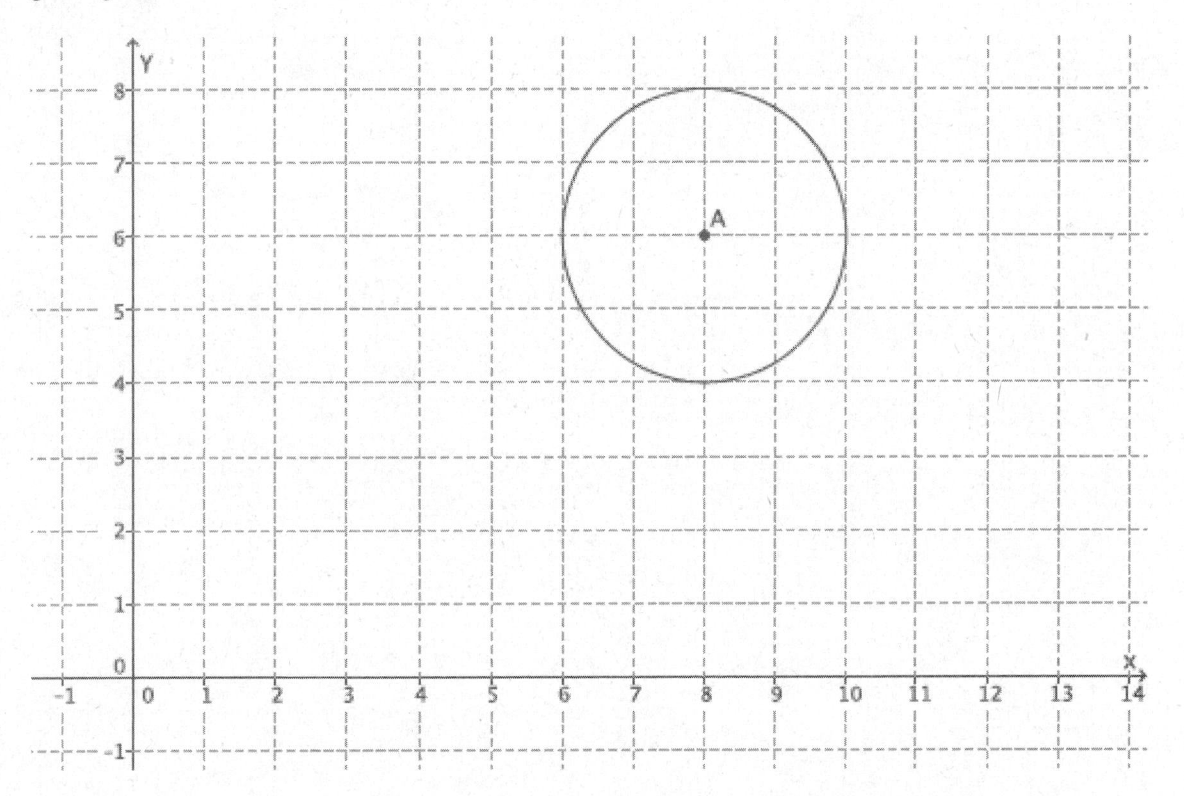

2. What scale factor would magnify the dilated circle back to the original size of circle A? How do you know?

©2019 Great Minds®. eureka-math.org

1. Dilate the figure from center O by a scale factor of $r = 3$. Make sure to use enough points to make a good image of the original figure.

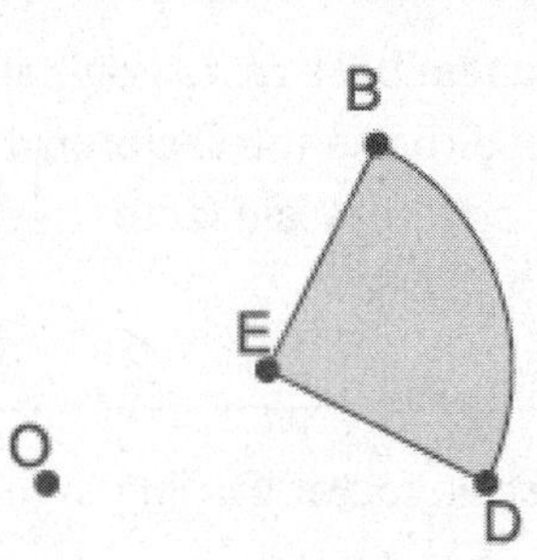

If I only dilate the points B, E, and D, then the dilated figure will look like a triangle. I will need to dilate more points along the curve BD.

The dilated image is shown in red below. Several points are needed along the curved portion of the diagram to produce an image similar to the original.

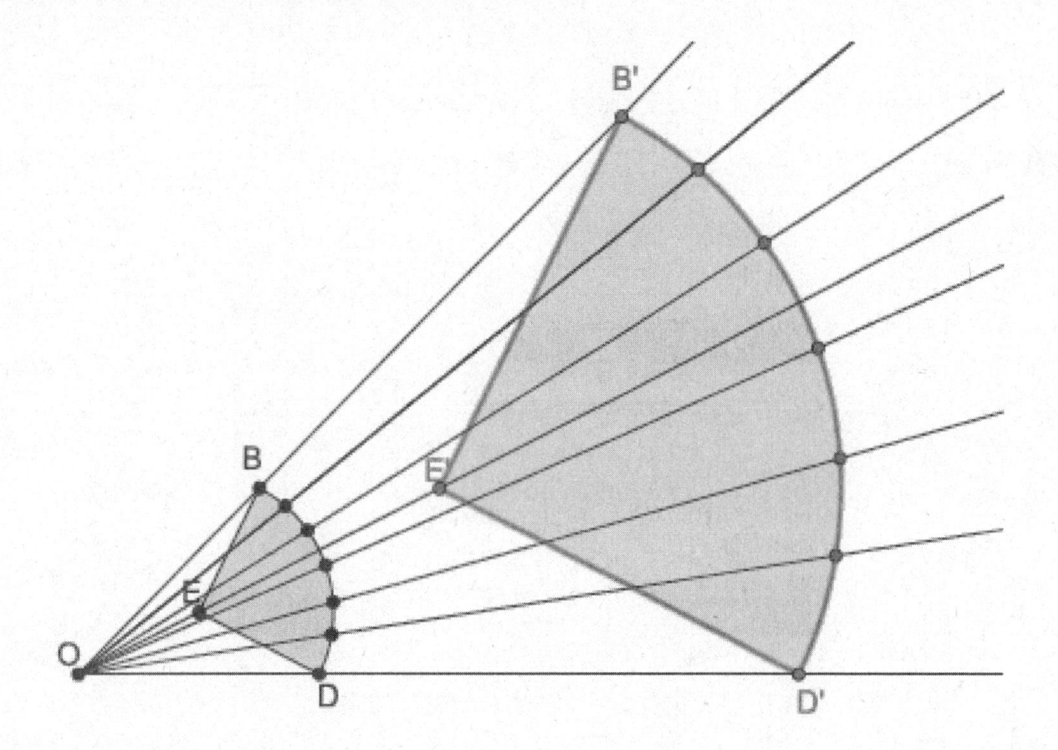

©2019 Great Minds®. eureka-math.org

2. A triangle ABC was dilated from center O by a scale factor of $r = 8$. What scale factor would shrink the dilated figure back to the original size?

A scale factor of $r = \frac{1}{8}$ would bring the dilated figure back to its original size.

To dilate back to its original size, I need to use the reciprocal of the original scale factor because $r \cdot \frac{1}{r} = 1$.

3. A figure has been dilated from center O by a scale factor of $r = \frac{2}{3}$. What scale factor would magnify the dilated figure back to the original size?

A scale factor of $r = \frac{3}{2}$ would bring the dilated figure back to its original size.

©2019 Great Minds®. eureka-math.org

1. Dilate the figure from center O by a scale factor $r = 2$. Make sure to use enough points to make a good image of the original figure.

O

2. Describe the process for selecting points when dilating a curved figure.

3. A figure was dilated from center O by a scale factor of $r = 5$. What scale factor would shrink the dilated figure back to the original size?

4. A figure has been dilated from center O by a scale factor of $r = \frac{7}{6}$. What scale factor would shrink the dilated figure back to the original size?

5. A figure has been dilated from center O by a scale factor of $r = \frac{3}{10}$. What scale factor would magnify the dilated figure back to the original size?

©2019 Great Minds®. eureka-math.org

Exercise

In the diagram below, points R and S have been dilated from center O by a scale factor of $r = 3$.

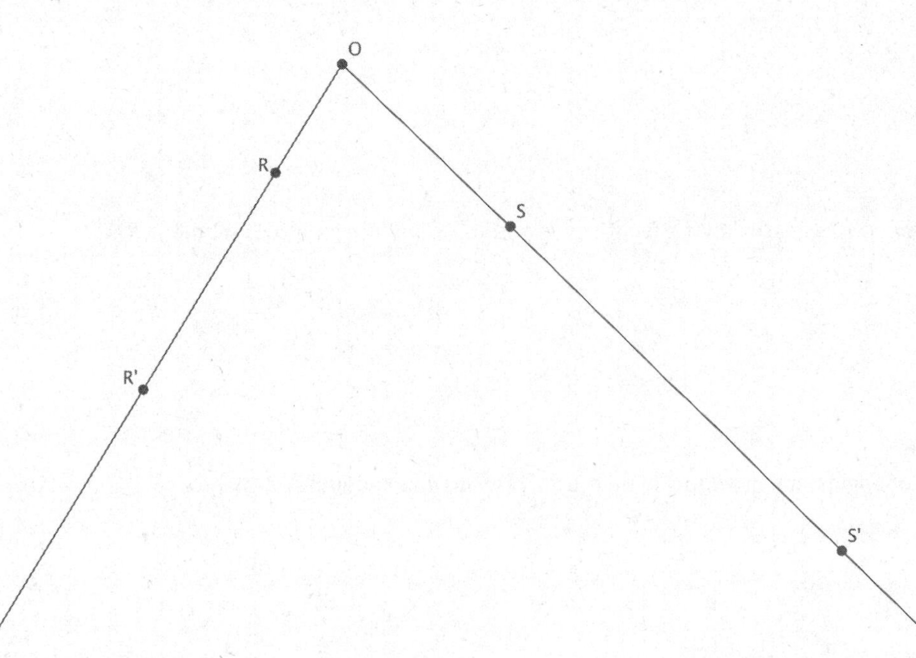

a. If $|OR| = 2.3 \text{ cm}$, what is $|OR'|$?

b. If $|OS| = 3.5 \text{ cm}$, what is $|OS'|$?

©2019 Great Minds®. eureka-math.org

c. Connect the point R to the point S and the point R' to the point S'. What do you know about the lines that contain segments RS and $R'S'$?

d. What is the relationship between the length of segment RS and the length of segment $R'S'$?

e. Identify pairs of angles that are equal in measure. How do you know they are equal?

©2019 Great Minds®. eureka-math.org

Lesson Summary

THEOREM: Given a dilation with center O and scale factor r, then for any two points P and Q in the plane so that O, P, and Q are not collinear, the lines PQ and $P'Q'$ are parallel, where $P' = Dilation(P)$ and $Q' = Dilation(Q)$, and furthermore, $|P'Q'| = r|PQ|$.

©2019 Great Minds®. eureka-math.org

Name ______________________________ Date ______________

Steven sketched the following diagram on graph paper. He dilated points B and C from point O. Answer the following questions based on his drawing.

1. What is the scale factor r? Show your work.

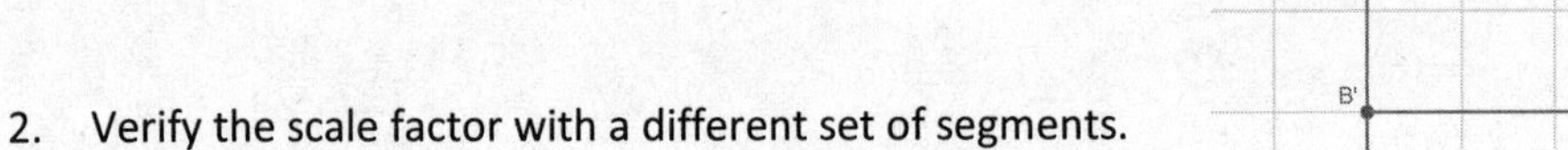

2. Verify the scale factor with a different set of segments.

3. Which segments are parallel? How do you know?

4. Are $\angle OBC$ and $\angle OB'C'$ right angles? How do you know?

©2019 Great Minds®. eureka-math.org

1. In the diagram below, points P, Q, and A have been dilated from center O by a scale factor of $r = \frac{1}{5}$ on a piece of lined paper.

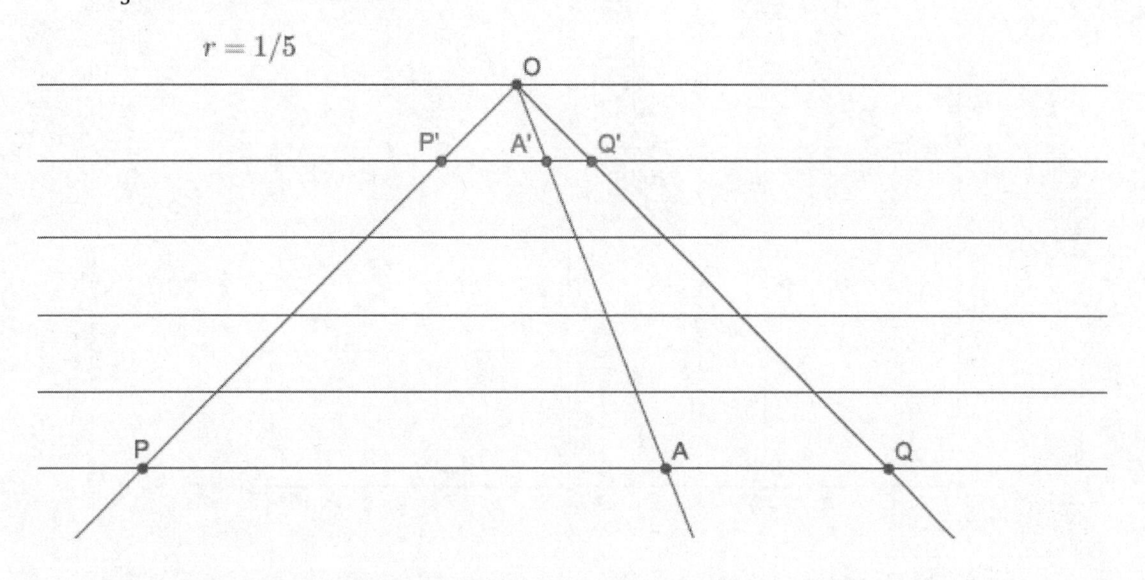

a. What is the relationship between segments PQ and $P'Q'$? How do you know?

$\overline{PQ}$ and $\overline{P'Q'}$ are parallel because they follow the lines on the lined paper.

b. What is the relationship between segments PA and $P'A'$? How do you know?

$\overline{PA}$ and $\overline{P'A'}$ are also parallel because they follow the lines on the lined paper.

c. Identify two angles whose measures are equal. How do you know?

$\angle OAP$ and $\angle OA'P'$ are equal in measure because they are corresponding angles created by parallel lines AP and $A'P'$.

d. What is the relationship between the lengths of segments AQ and $A'Q'$? How do you know?

The length of segment $A'Q'$ will be $\frac{1}{5}$ the length of segment AQ. The FTS states that the length of the dilated segment is equal to the scale factor multiplied by the original segment length, or $|A'Q'| = r|AQ|$.

©2019 Great Minds®. eureka-math.org

2. Reynaldo sketched the following diagram on graph paper. He dilated points B and C from center O.

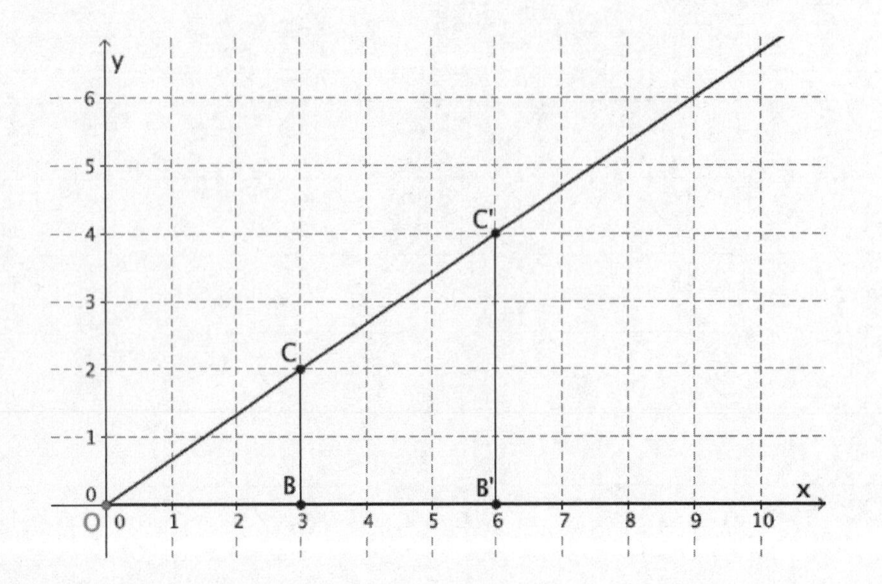

a. What is the scale factor r? Show your work.

$$|OB'| = r|OB|$$
$$6 = r(3)$$
$$\frac{6}{3} = r$$
$$2 = r$$

b. Verify the scale factor with a different set of segments.

$$|B'C'| = r|BC|$$
$$4 = r(2)$$
$$\frac{4}{2} = r$$
$$2 = r$$

c. Which segments are parallel? How do you know?

Segments BC and $B'C'$ are parallel. They are on the lines of the grid paper, which I know are parallel.

d. Which angles are equal in measure? How do you know?

$|\angle OB'C'| = |\angle OBC|$, and $|\angle OC'B'| = |\angle OCB|$ because they are corresponding angles of parallel lines cut by a transversal.

©2019 Great Minds®. eureka-math.org

3. Points B and C were dilated from center O.

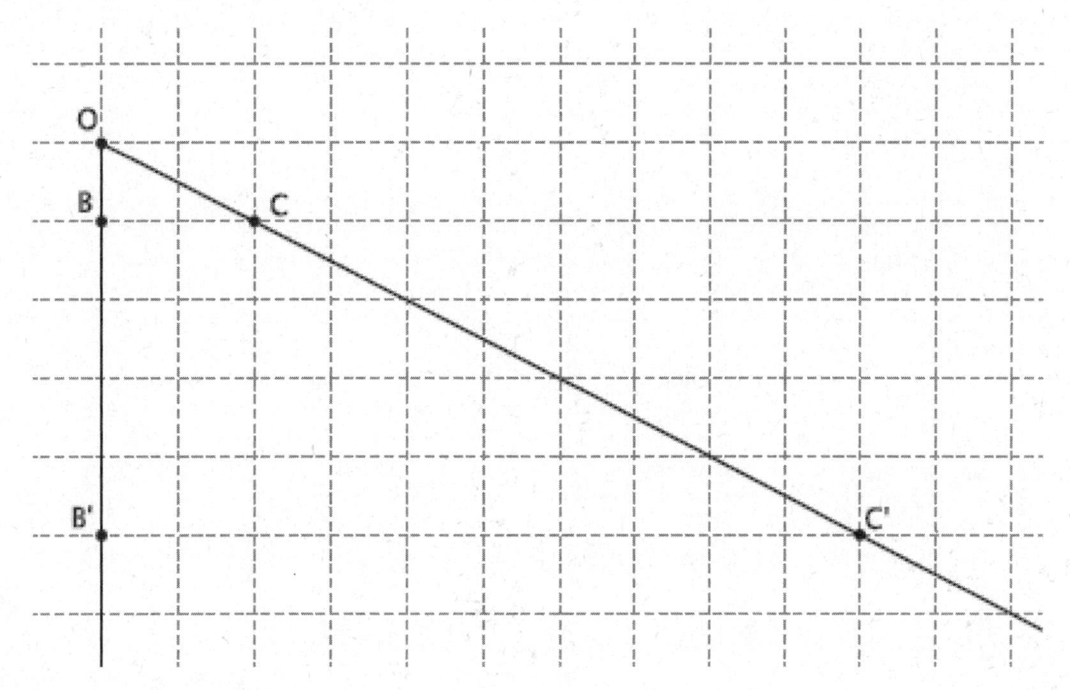

a. What is the scale factor r? Show your work.

$$|OB'| = r|OB|$$
$$5 = r(1)$$
$$5 = r$$

b. If $|OC| \approx 2.2$, what is $|OC'|$?

$$|OC'| = r|OC|$$
$$|OC'| \approx 5(2.2)$$
$$|OC'| \approx 11$$

c. How does the perimeter of $\triangle OBC$ compare to the perimeter of $\triangle OB'C'$?

Perimeter $\triangle OBC \approx 1 + 2 + 2.2$

Perimeter $\triangle OBC \approx 5.2$

Perimeter $\triangle OB'C' \approx 5 + 10 + 11$

Perimeter $\triangle OB'C' \approx 26$

d. Was the perimeter of $\triangle OB'C'$ equal to the perimeter of $\triangle OBC$ multiplied by scale factor r? Explain.

Yes. The perimeter of $\triangle OB'C'$ ***was five times the perimeter of*** $\triangle OBC$***, which makes sense because the dilation increased the length of each segment by a scale factor of*** 5***. That means that each side of*** $\triangle OB'C'$ ***was five times as long as each side of*** $\triangle OBC$***.***

©2019 Great Minds®. eureka-math.org

1. Use a piece of notebook paper to verify the fundamental theorem of similarity for a scale factor r that is $0 < r < 1$.

   ✓ Mark a point O on the first line of notebook paper.

   ✓ Mark the point P on a line several lines down from the center O. Draw a ray, $\overrightarrow{OP}$. Mark the point P' on the ray and on a line of the notebook paper closer to O than you placed point P. This ensures that you have a scale factor that is $0 < r < 1$. Write your scale factor at the top of the notebook paper.

   ✓ Draw another ray, $\overrightarrow{OQ}$, and mark the points Q and Q' according to your scale factor.

   ✓ Connect points P and Q. Then, connect points P' and Q'.

   ✓ Place a point, A, on the line containing segment PQ between points P and Q. Draw ray $\overrightarrow{OA}$. Mark point A' at the intersection of the line containing segment $P'Q'$ and ray $\overrightarrow{OA}$.

   a. Are the lines containing segments PQ and $P'Q'$ parallel lines? How do you know?

   b. Which, if any, of the following pairs of angles are equal in measure? Explain.

      i. $\angle OPQ$ and $\angle OP'Q'$

      ii. $\angle OAQ$ and $\angle OA'Q'$

      iii. $\angle OAP$ and $\angle OA'P'$

      iv. $\angle OQP$ and $\angle OQ'P'$

   c. Which, if any, of the following statements are true? Show your work to verify or dispute each statement.

      i. $|OP'| = r|OP|$

      ii. $|OQ'| = r|OQ|$

      iii. $|P'A'| = r|PA|$

      iv. $|A'Q'| = r|AQ|$

   d. Do you believe that the fundamental theorem of similarity (FTS) is true even when the scale factor is $0 < r < 1$? Explain.

©2019 Great Minds®. eureka-math.org

2. Caleb sketched the following diagram on graph paper. He dilated points B and C from center O.

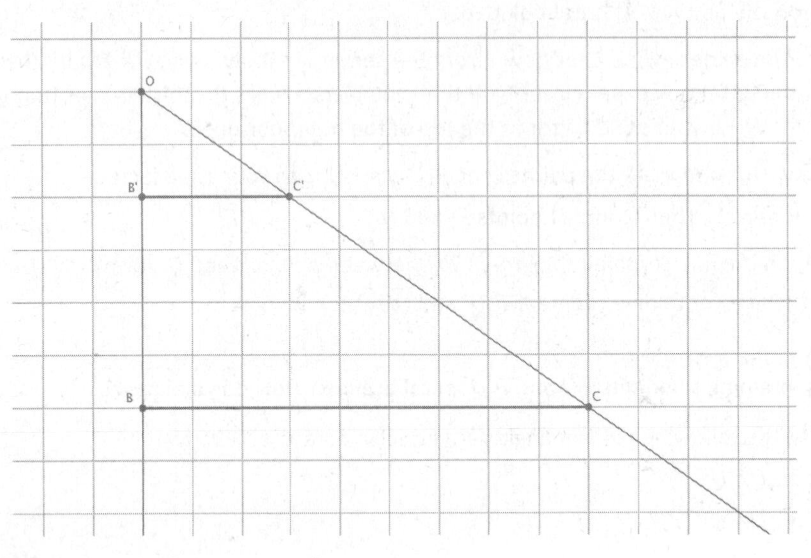

a. What is the scale factor r? Show your work.

b. Verify the scale factor with a different set of segments.

c. Which segments are parallel? How do you know?

d. Which angles are equal in measure? How do you know?

3. Points B and C were dilated from center O.

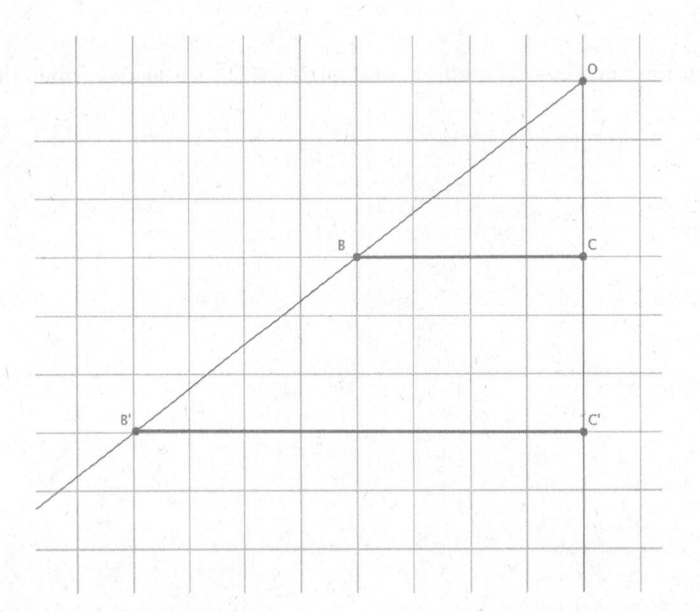

a. What is the scale factor r? Show your work.

b. If $|OB| = 5$, what is $|OB'|$?

c. How does the perimeter of triangle OBC compare to the perimeter of triangle $OB'C'$?

d. Did the perimeter of triangle $OB'C' = r \times$ (perimeter of triangle OBC)? Explain.

©2019 Great Minds®. eureka-math.org

Exercise 1

In the diagram below, points P and Q have been dilated from center O by scale factor r. $\overline{PQ} \parallel \overline{P'Q'}$, $|PQ| = 5 \text{ cm}$, and $|P'Q'| = 10 \text{ cm}$.

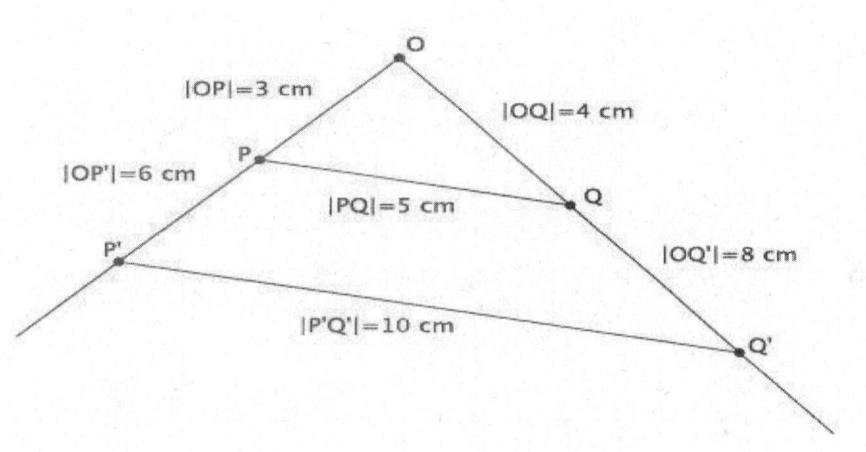

a. Determine the scale factor r.

b. Locate the center O of dilation. Measure the segments to verify that $|OP'| = r|OP|$ and $|OQ'| = r|OQ|$. Show your work below.

©2019 Great Minds®. eureka-math.org

Exercise 2

In the diagram below, you are given center O and ray $\overrightarrow{OA}$. Point A is dilated by a scale factor $r = 4$. Use what you know about FTS to find the location of point A'.

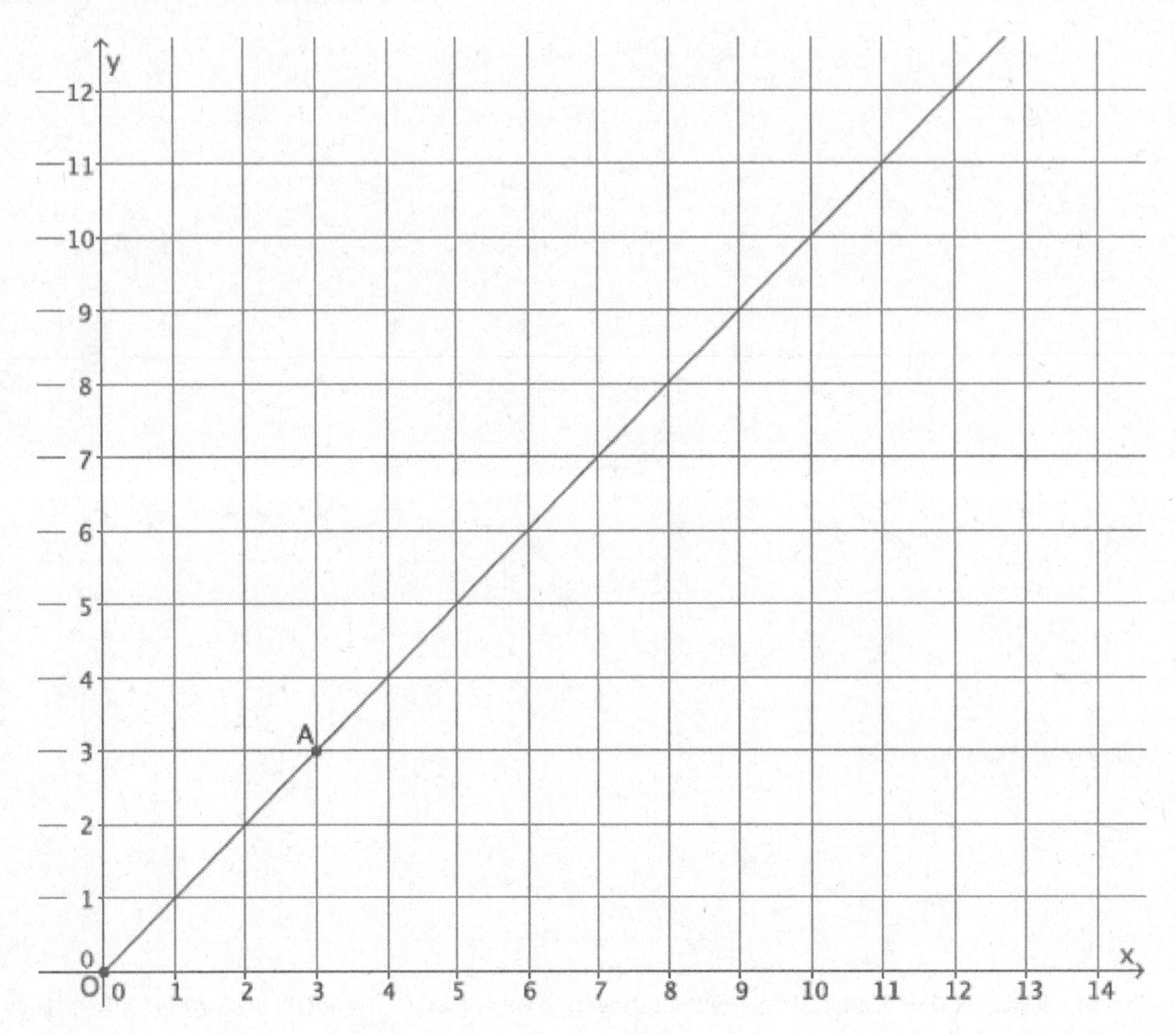

©2019 Great Minds®. eureka-math.org

Exercise 3

In the diagram below, you are given center O and ray $\overrightarrow{OA}$. Point A is dilated by a scale factor $r = \frac{5}{12}$. Use what you know about FTS to find the location of point A'.

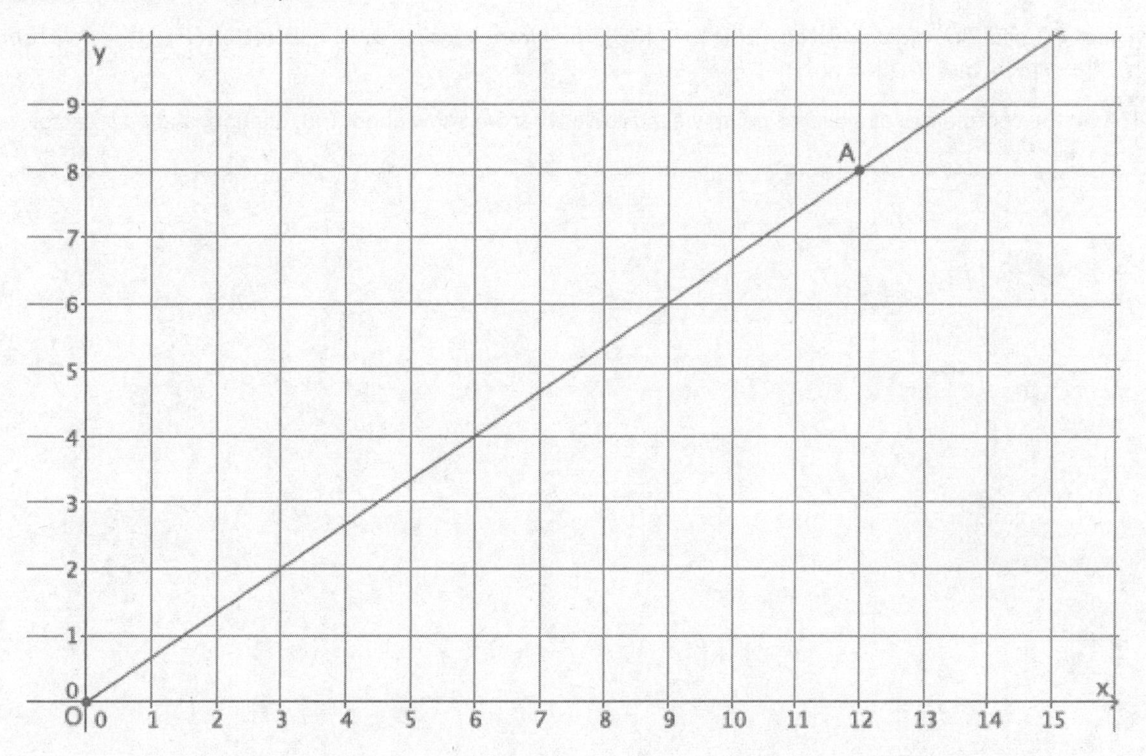

©2019 Great Minds®. eureka-math.org

Lesson Summary

Converse of the fundamental theorem of similarity:

If lines PQ *and* $P'Q'$ *are parallel and* $|P'Q'| = r|PQ|$, *then from a center* O, $P' = Dilation(P)$, $Q' = Dilation(Q)$, $|OP'| = r|OP|$, *and* $|OQ'| = r|OQ|$.

To find the coordinates of a dilated point, we must use what we know about FTS, dilation, and scale factor.

©2019 Great Minds®. eureka-math.org

Name ____ Date ____________________

In the diagram below, you are given center O and ray $\overrightarrow{OA}$. Point A is dilated by a scale factor $r = \frac{6}{4}$. Use what you know about FTS to find the location of point A'.

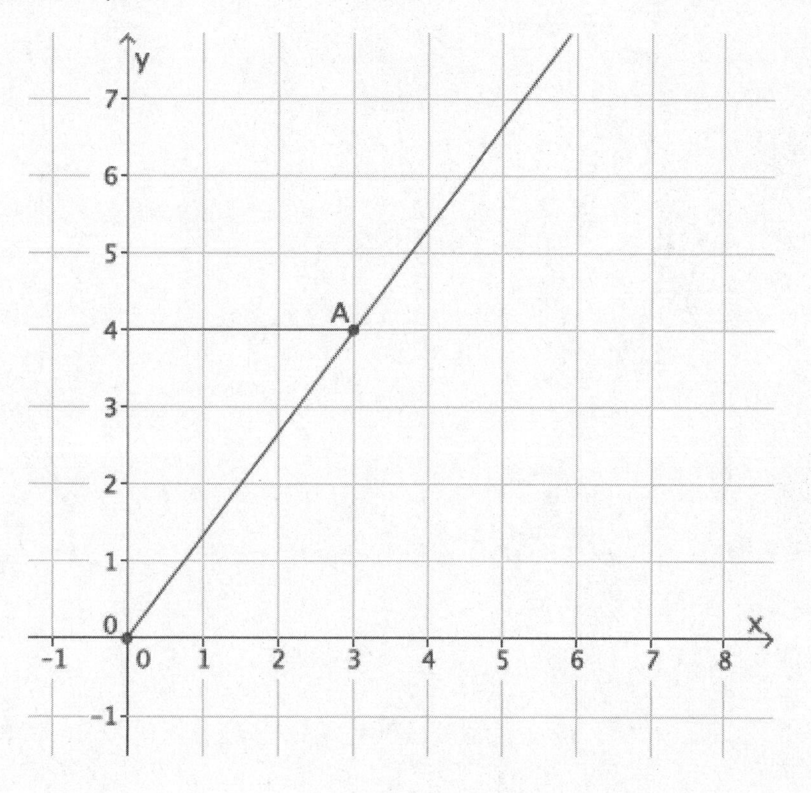

©2019 Great Minds®. eureka-math.org

1. Dilate point A, located at $(2, 4)$ from center O, by a scale factor of $r = \frac{7}{2}$. What is the precise location of point A'?

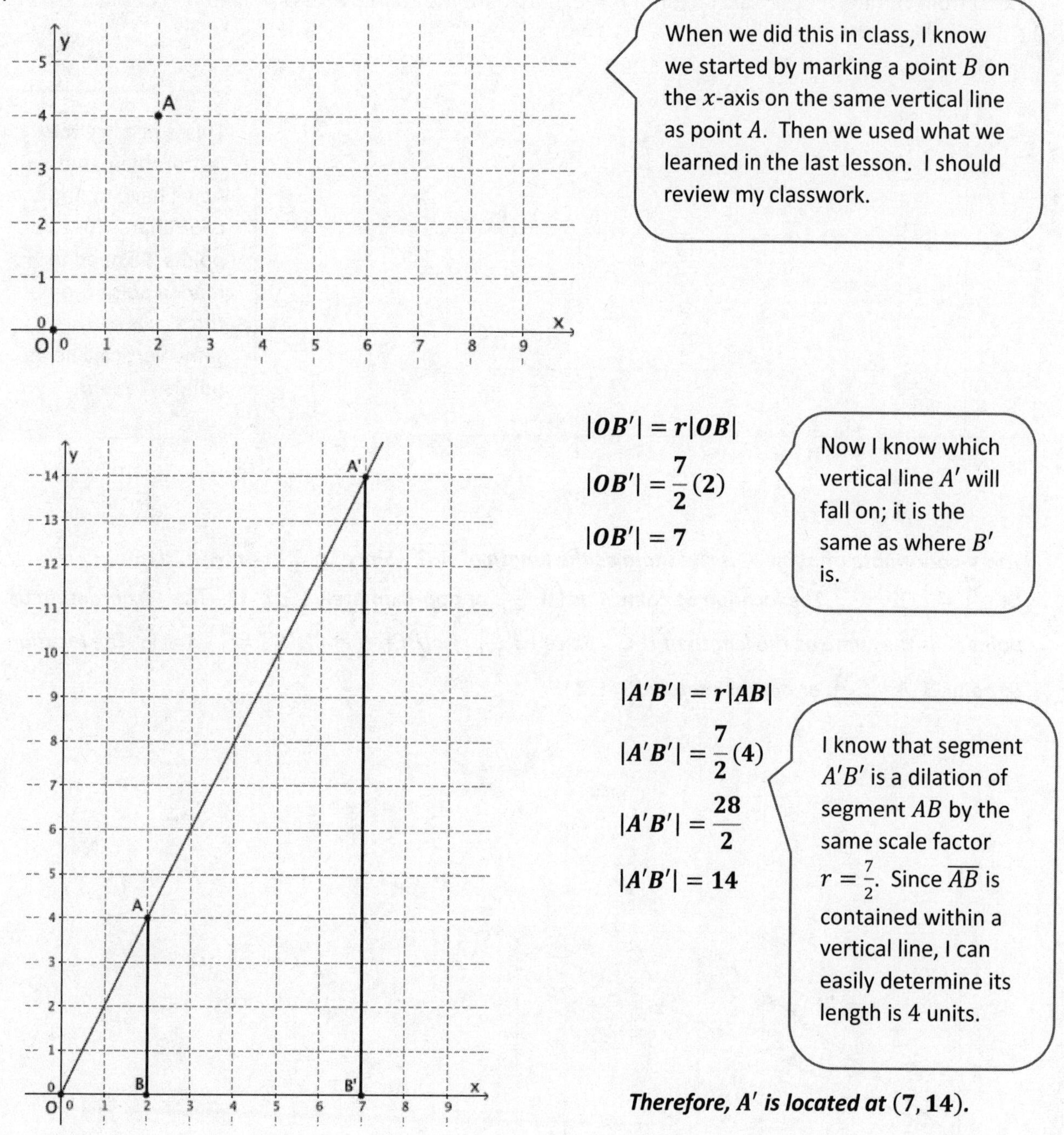

©2019 Great Minds®. eureka-math.org

2. Dilate point A, located at $(7, 5)$ from center O, by a scale factor of $r = \frac{3}{7}$. Then, dilate point B, located at $(7, 3)$ from center O, by a scale factor of $r = \frac{3}{7}$. What are the coordinates of A' and B'? Explain.

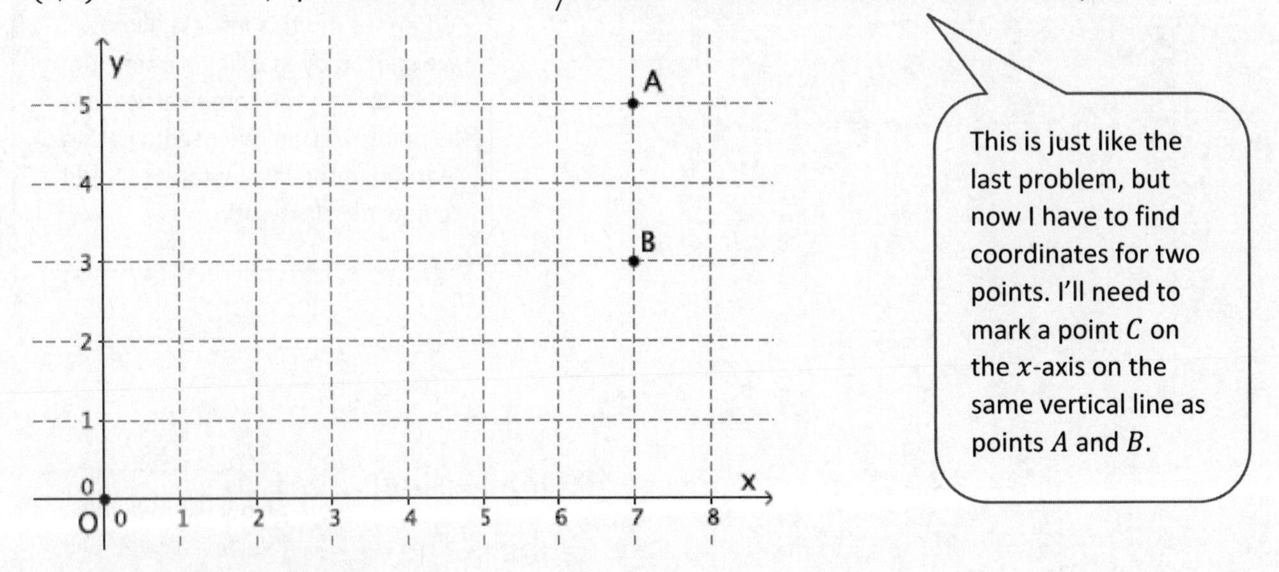

The y-coordinate of point A' is the same as the length of $\overline{A'C'}$. Since $|A'C'| = r|AC|$, then $|A'C'| = \frac{3}{7} \cdot 5 = \frac{15}{7}$. The location of point A' is $\left(3, \frac{15}{7}\right)$, or approximately $(3, 2.1)$. The y-coordinate of point B' is the same as the length of $\overline{B'C'}$. Since $|B'C'| = r|BC|$, then $|B'C'| = \frac{3}{7} \cdot 3 = \frac{9}{7}$. The location of point B' is $\left(3, \frac{9}{7}\right)$, or approximately $(3, 1.3)$.

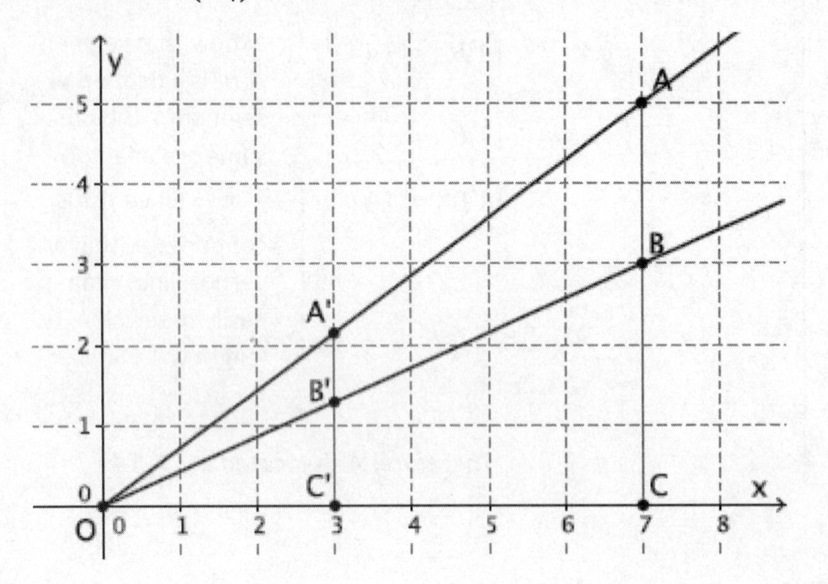

©2019 Great Minds®. eureka-math.org

1. Dilate point A, located at $(3, 4)$ from center O, by a scale factor $r = \frac{5}{3}$.

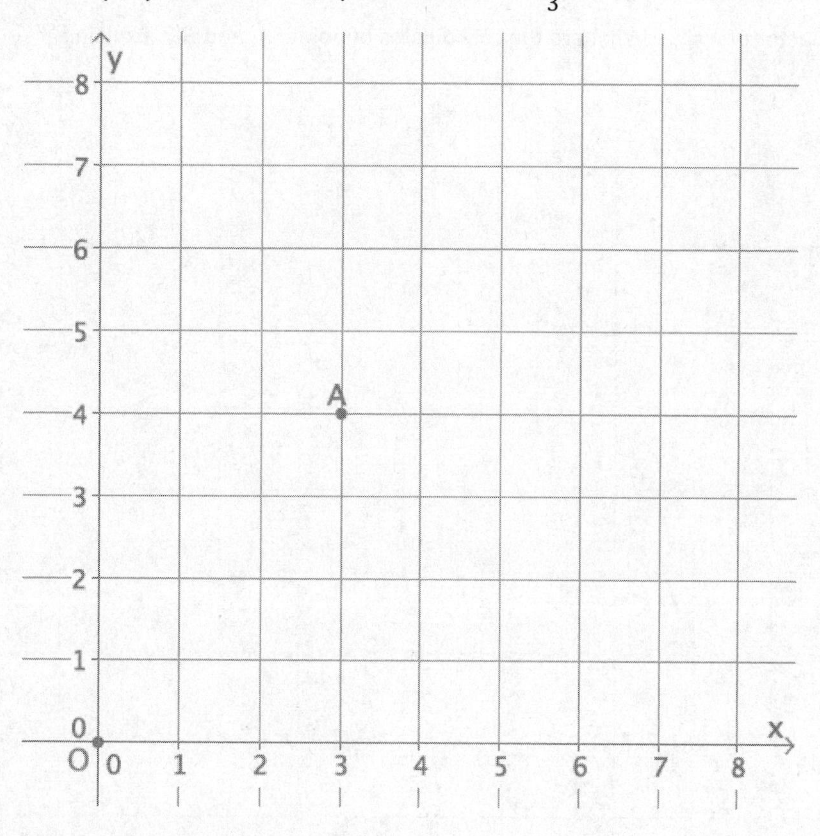

What is the precise location of point A'?

©2019 Great Minds®. eureka-math.org

2. Dilate point A, located at $(9, 7)$ from center O, by a scale factor $r = \frac{4}{9}$. Then, dilate point B, located at $(9, 5)$ from center O, by a scale factor of $r = \frac{4}{9}$. What are the coordinates of points A' and B'? Explain.

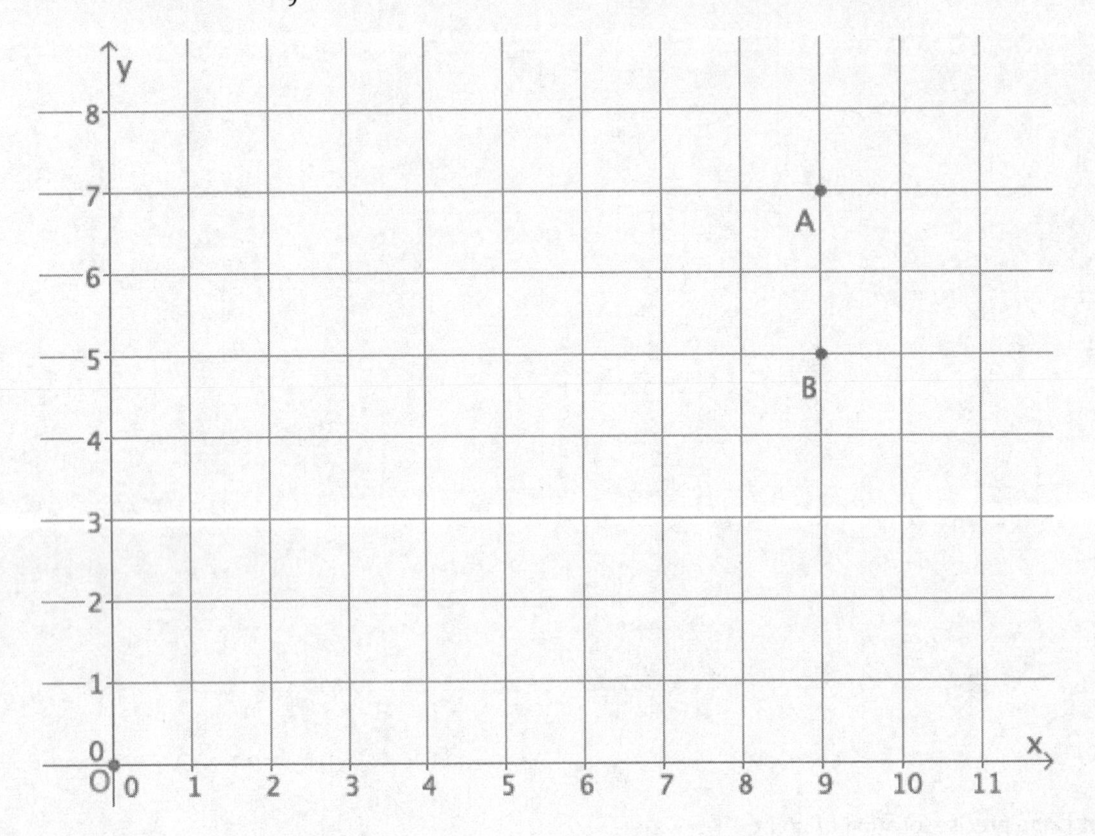

3. Explain how you used the fundamental theorem of similarity in Problems 1 and 2.

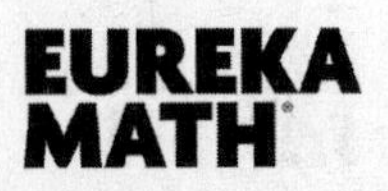

©2019 Great Minds®. eureka-math.org

Exercises 1–5

1. Point $A(7, 9)$ is dilated from the origin by scale factor $r = 6$. What are the coordinates of point A'?

2. Point $B(-8, 5)$ is dilated from the origin by scale factor $r = \frac{1}{2}$. What are the coordinates of point B'?

3. Point $C(6, -2)$ is dilated from the origin by scale factor $r = \frac{3}{4}$. What are the coordinates of point C'?

4. Point $D(0, 11)$ is dilated from the origin by scale factor $r = 4$. What are the coordinates of point D'?

5. Point $E(-2, -5)$ is dilated from the origin by scale factor $r = \frac{3}{2}$. What are the coordinates of point E'?

©2019 Great Minds®. eureka-math.org

Exercises 6–8

6. The coordinates of triangle ABC are shown on the coordinate plane below. The triangle is dilated from the origin by scale factor $r = 12$. Identify the coordinates of the dilated triangle $A'B'C'$.

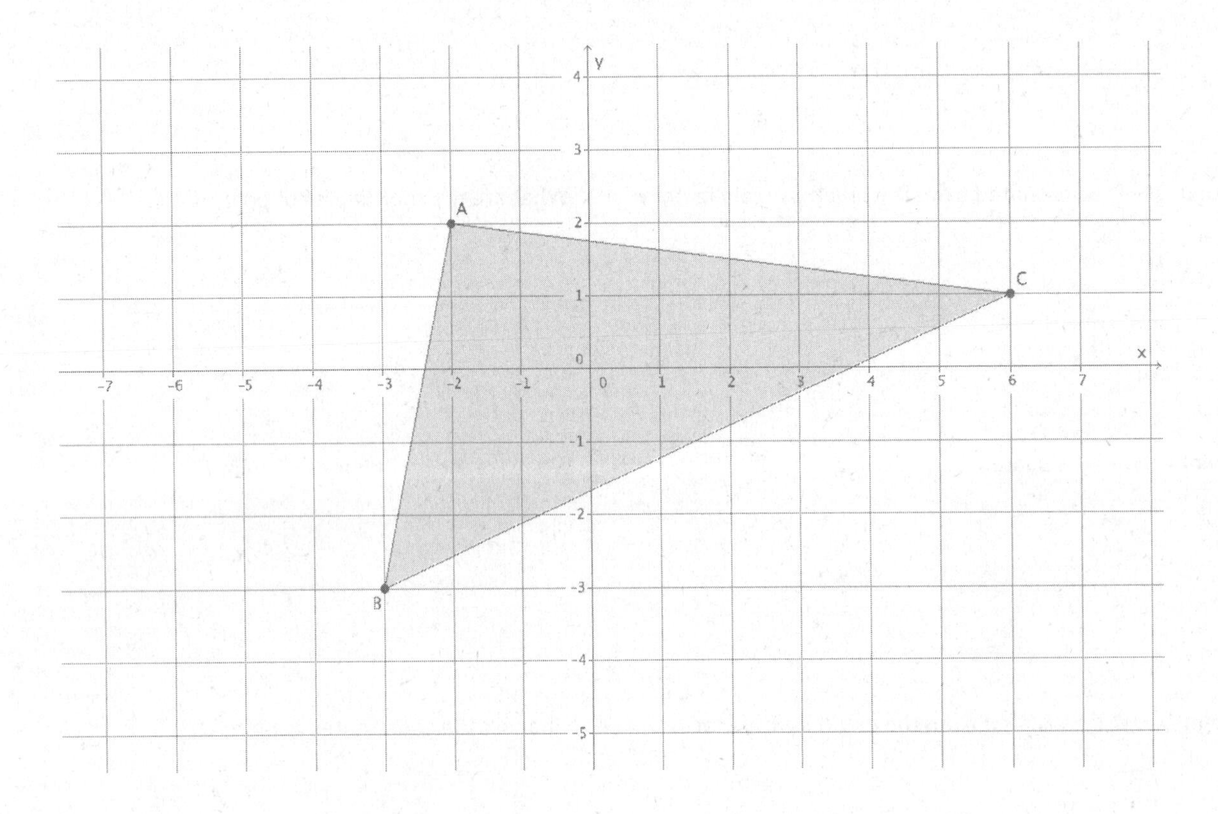

©2019 Great Minds®. eureka-math.org

7. Figure $DEFG$ is shown on the coordinate plane below. The figure is dilated from the origin by scale factor $r = \frac{2}{3}$. Identify the coordinates of the dilated figure $D'E'F'G'$, and then draw and label figure $D'E'F'G'$ on the coordinate plane.

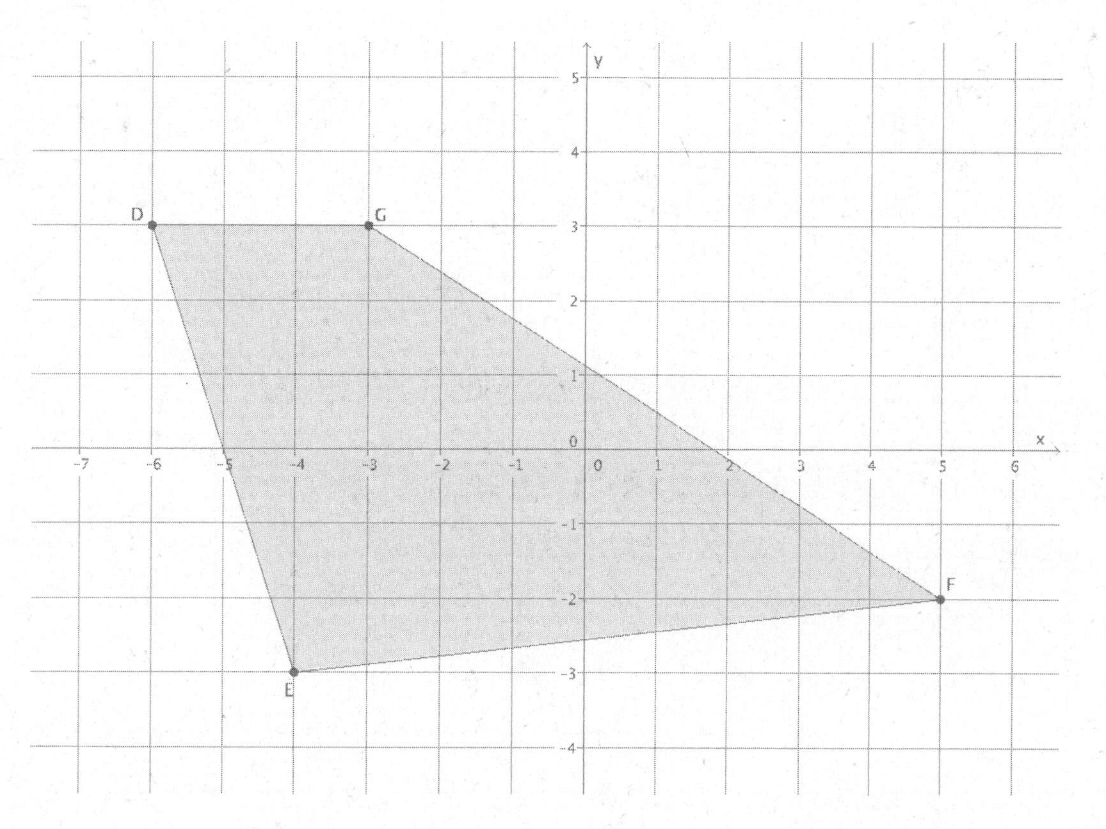

©2019 Great Minds®. eureka-math.org

8. The triangle ABC has coordinates $A(3, 2)$, $B(12, 3)$, and $C(9, 12)$. Draw and label triangle ABC on the coordinate plane. The triangle is dilated from the origin by scale factor $r = \frac{1}{3}$. Identify the coordinates of the dilated triangle $A'B'C'$, and then draw and label triangle $A'B'C'$ on the coordinate plane.

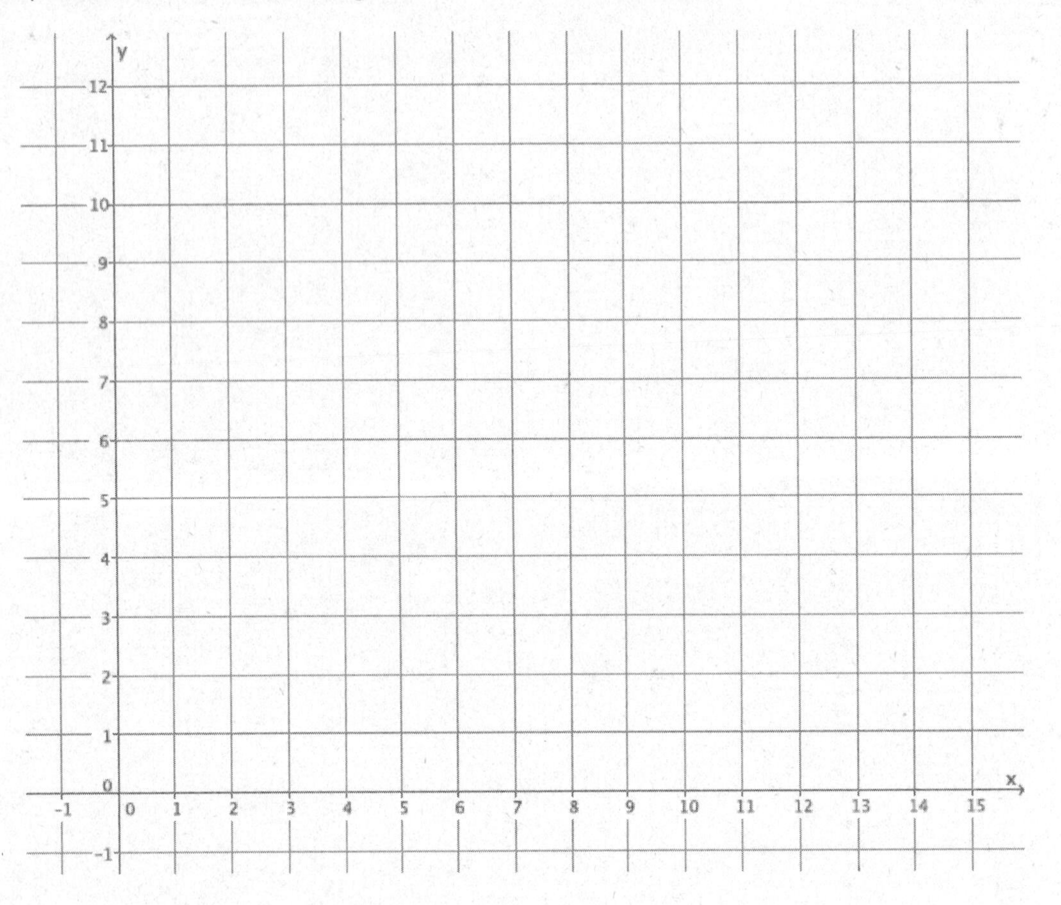

©2019 Great Minds®. eureka-math.org

Lesson Summary

Dilation has a multiplicative effect on the coordinates of a point in the plane. Given a point (x, y) in the plane, a dilation from the origin with scale factor r moves the point (x, y) to (rx, ry).

For example, if a point $(3, -5)$ in the plane is dilated from the origin by a scale factor of $r = 4$, then the coordinates of the dilated point are $\big(4 \cdot 3, 4 \cdot (-5)\big) = (12, -20)$.

©2019 Great Minds®. eureka-math.org

Name ____ Date ____________________

1. The point $A(7, 4)$ is dilated from the origin by a scale factor $r = 3$. What are the coordinates of point A'?

2. The triangle ABC, shown on the coordinate plane below, is dilated from the origin by scale factor $r = \frac{1}{2}$. What is the location of triangle $A'B'C'$? Draw and label it on the coordinate plane.

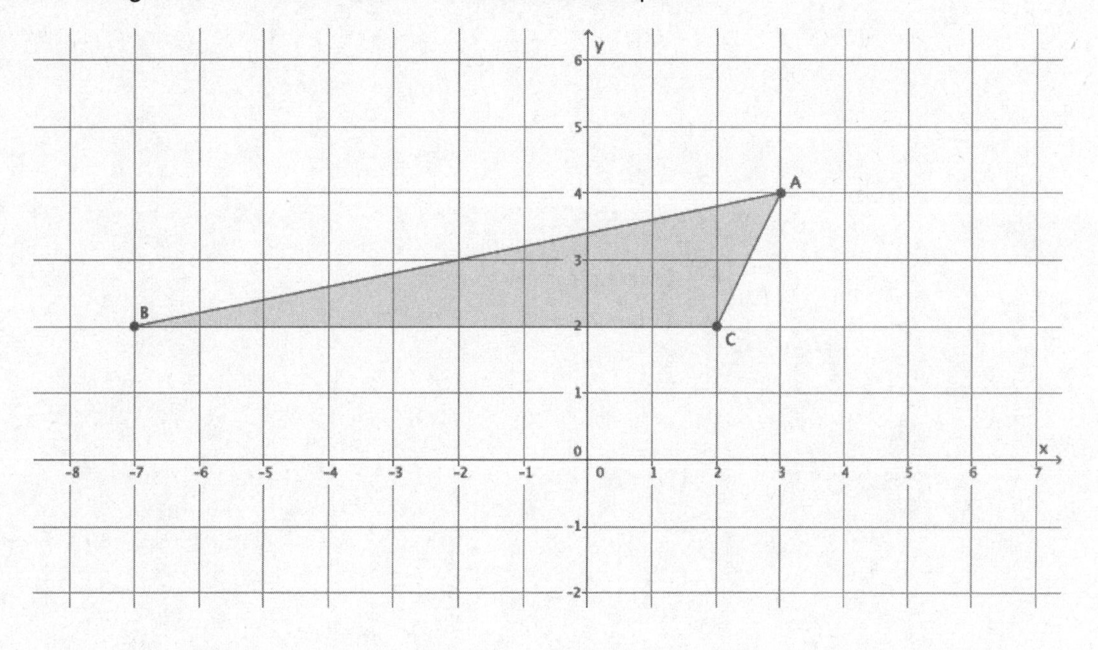

©2019 Great Minds®. eureka-math.org

1. Triangle ABC is shown on the coordinate plane below. The triangle is dilated from the origin by scale factor $r = 2$. Identify the coordinates of the dilated triangle $A'B'C'$.

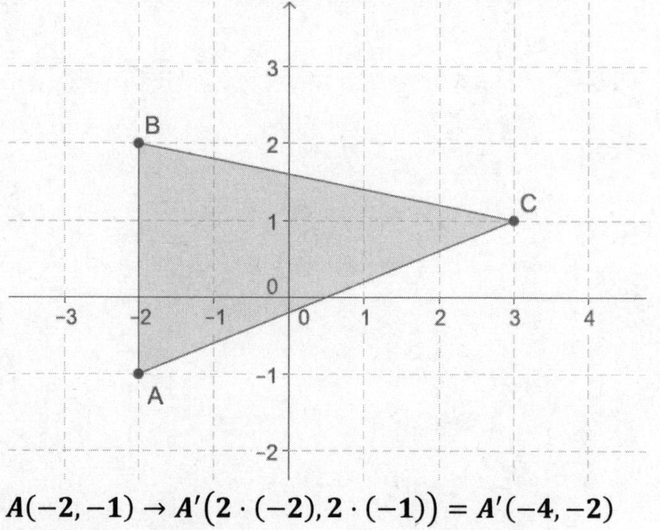

The work we did in class led us to the conclusion that when given a point $A(x, y)$, we can find the coordinates of A' using the scale factor: $A'(rx, ry)$. This only works for dilations from the origin.

$$\boldsymbol{A(-2,-1) \rightarrow A'\big(2 \cdot (-2), 2 \cdot (-1)\big) = A'(-4,-2)}$$
$$\boldsymbol{B(-2,2) \rightarrow B'(2 \cdot (-2), 2 \cdot 2) = B'(-4,4)}$$
$$\boldsymbol{C(3,1) \rightarrow C'(2 \cdot 3, 2 \cdot 1) = C'(6,2)}$$

The coordinates of the dilated triangle will be $A'(-4,-2)$, $B'(-4,4)$, $C'(6,2)$.

2. The figure $ABCD$ has coordinates $A(3,1)$, $B(12,9)$, $C(-9,3)$, and $D(-12,-3)$. The figure is dilated from the origin by a scale factor $r = \frac{2}{3}$. Identify the coordinates of the dilated figure $A'B'C'D'$.

$$\boldsymbol{A(3,1) \rightarrow A'\left(\frac{2}{3} \cdot 3, \frac{2}{3} \cdot 1\right) = A'\left(2, \frac{2}{3}\right)}$$
$$\boldsymbol{B(12,9) \rightarrow B'\left(\frac{2}{3} \cdot 12, \frac{2}{3} \cdot 9\right) = B'(8,6)}$$
$$\boldsymbol{C(-9,3) \rightarrow C'\left(\frac{2}{3} \cdot (-9), \frac{2}{3} \cdot 3\right) = C'(-6,2)}$$
$$\boldsymbol{D(-12,-3) \rightarrow D'\left(\frac{2}{3} \cdot (-12), \frac{2}{3} \cdot (-3)\right) = D'(-8,-2)}$$

The coordinates of the dilated figure are $A'\left(2, \frac{2}{3}\right)$, $B'(8,6)$, $C'(-6, 2)$, and $D'(-8,-2)$.

©2019 Great Minds®. eureka-math.org

1. Triangle ABC is shown on the coordinate plane below. The triangle is dilated from the origin by scale factor $r = 4$. Identify the coordinates of the dilated triangle $A'B'C'$.

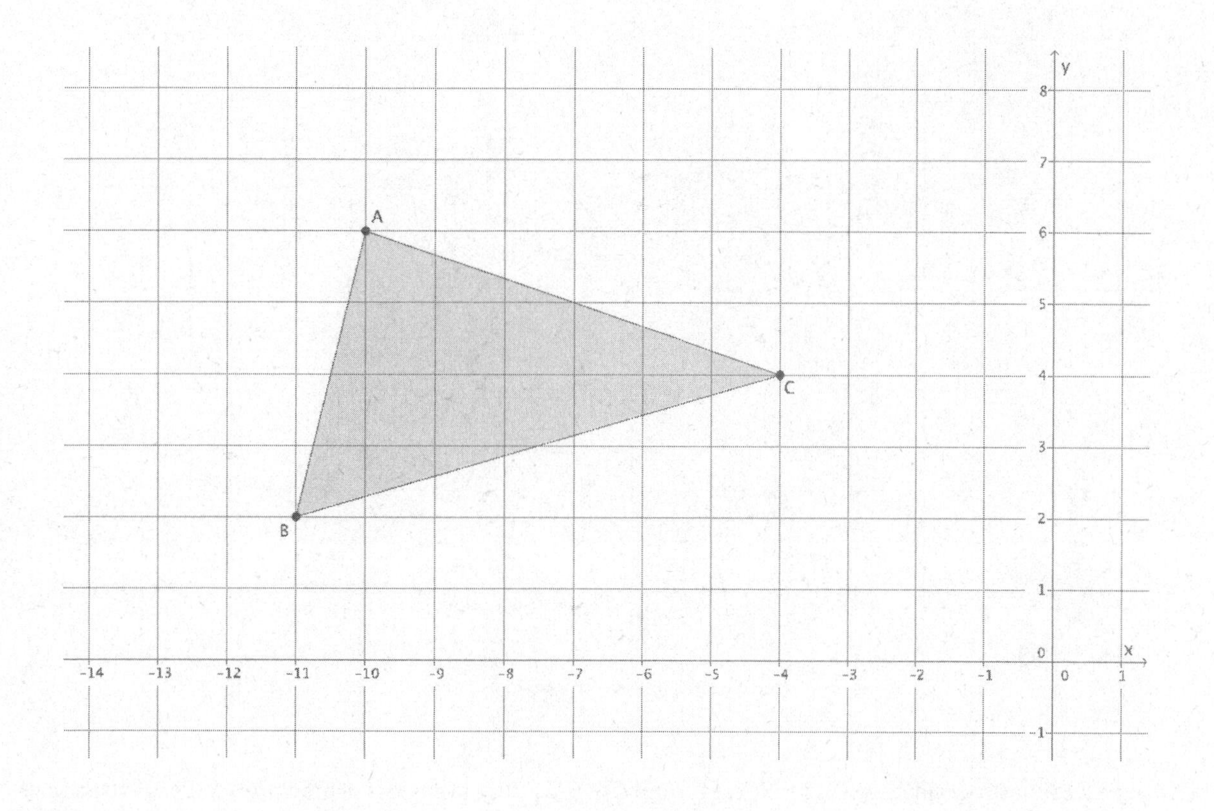

©2019 Great Minds®. eureka-math.org

2. Triangle ABC is shown on the coordinate plane below. The triangle is dilated from the origin by scale factor $r = \frac{5}{4}$. Identify the coordinates of the dilated triangle $A'B'C'$.

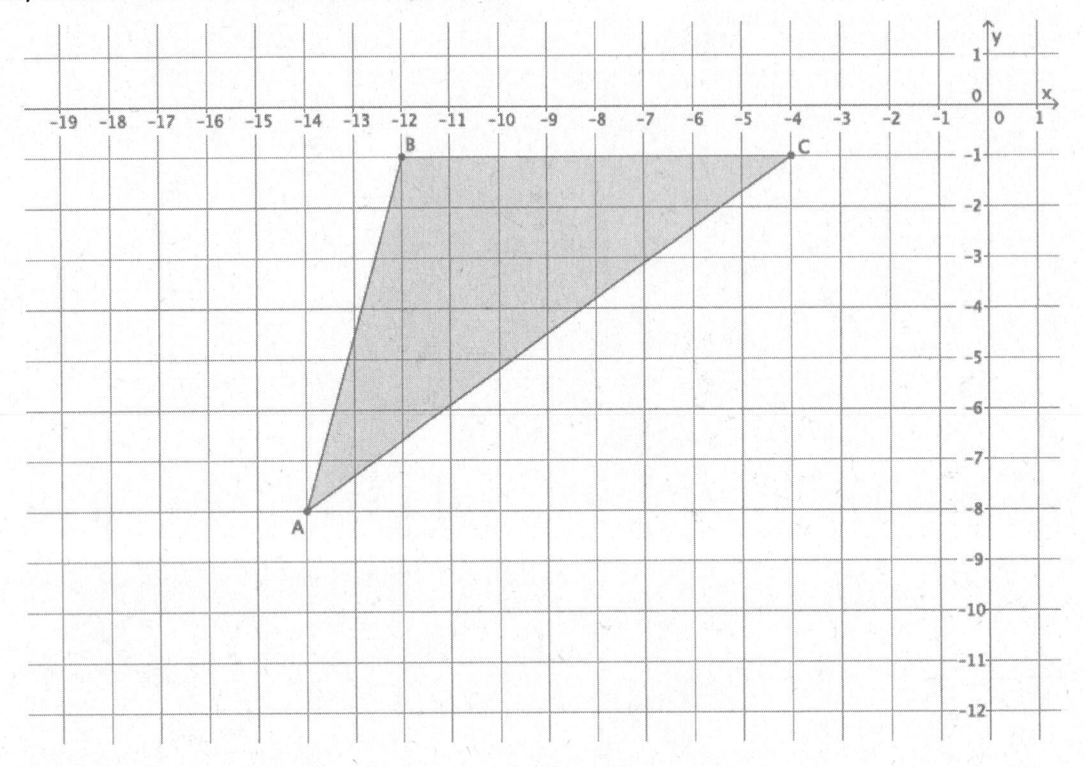

3. The triangle ABC has coordinates $A(6, 1)$, $B(12, 4)$, and $C(-6, 2)$. The triangle is dilated from the origin by a scale factor $r = \frac{1}{2}$. Identify the coordinates of the dilated triangle $A'B'C'$.

©2019 Great Minds®. eureka-math.org

4. Figure $DEFG$ is shown on the coordinate plane below. The figure is dilated from the origin by scale factor $r = \frac{3}{2}$. Identify the coordinates of the dilated figure $D'E'F'G'$, and then draw and label figure $D'E'F'G'$ on the coordinate plane.

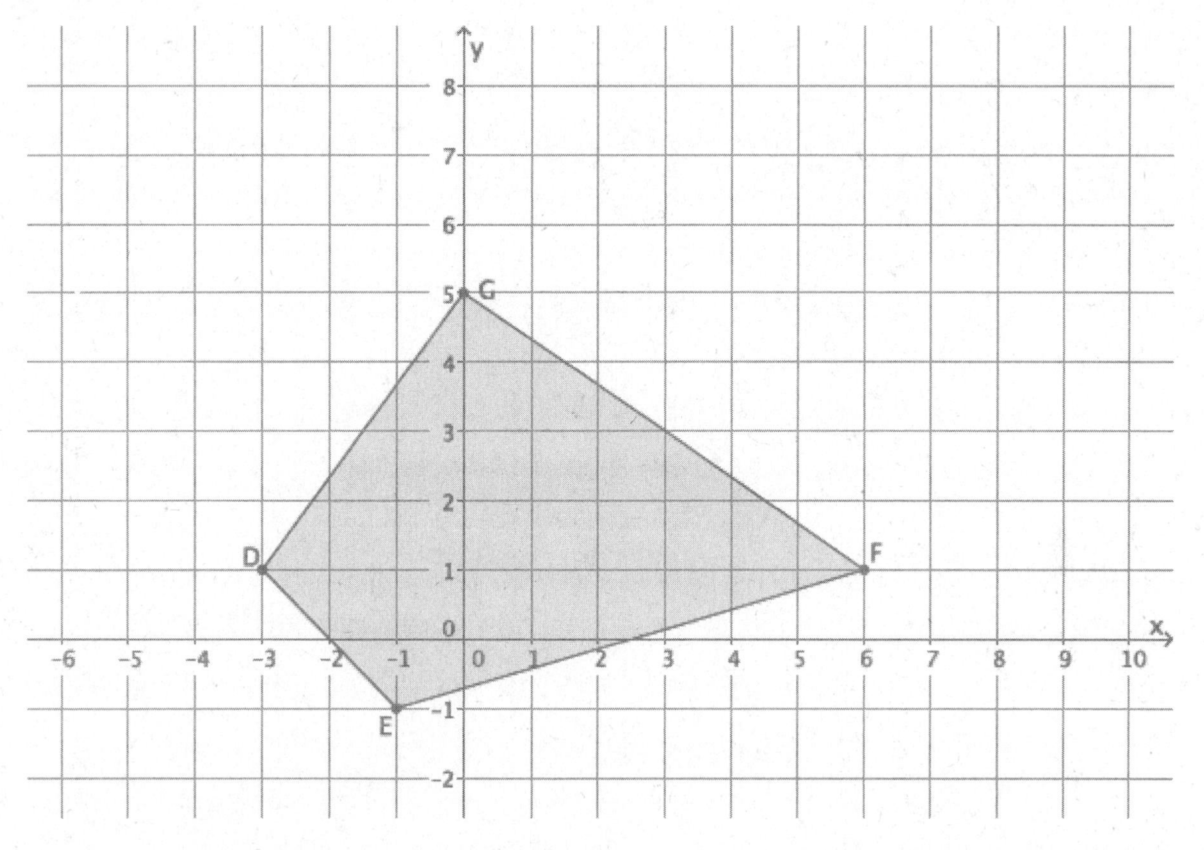

5. Figure $DEFG$ has coordinates $D(1, 1)$, $E(7, 3)$, $F(5, -4)$, and $G(-1, -4)$. The figure is dilated from the origin by scale factor $r = 7$. Identify the coordinates of the dilated figure $D'E'F'G'$.

©2019 Great Minds®. eureka-math.org

Exercise

Use the diagram below to prove the theorem: *Dilations preserve the measures of angles.*

Let there be a dilation from center O with scale factor r. Given $\angle PQR$, show that since $P' = Dilation(P)$, $Q' = Dilation(Q)$, and $R' = Dilation(R)$, then $|\angle PQR| = |\angle P'Q'R'|$. That is, show that the image of the angle after a dilation has the same measure, in degrees, as the original.

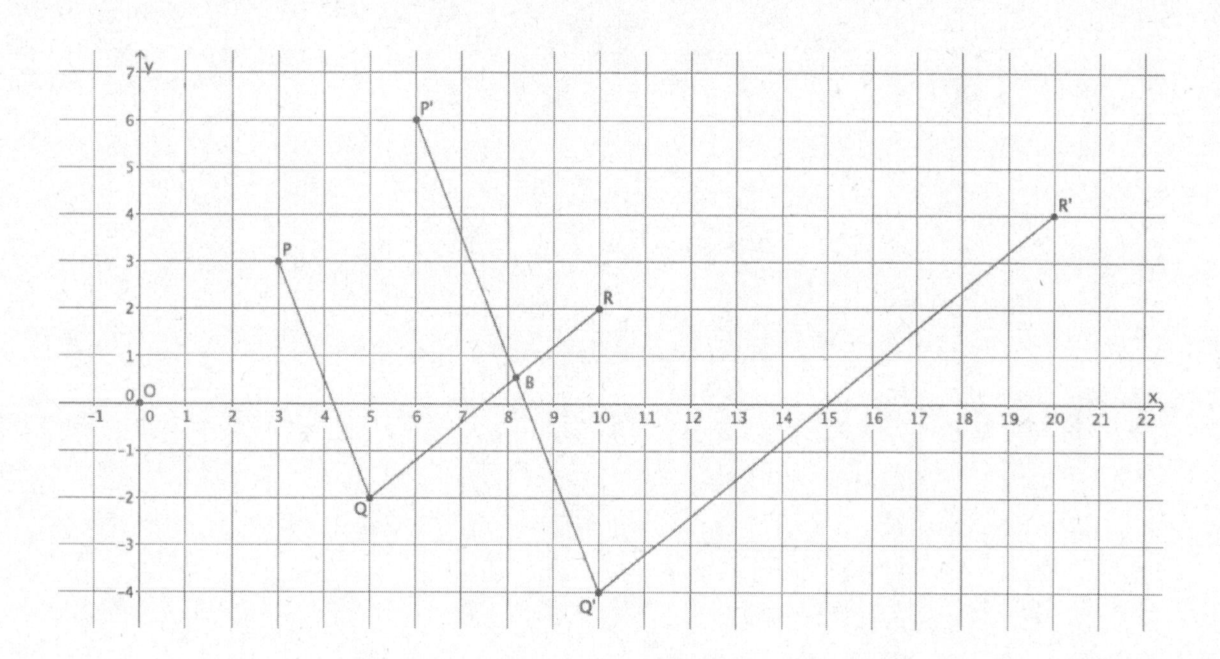

©2019 Great Minds®. eureka-math.org

Name ______________________________ Date ______________

Dilate $\angle ABC$ with center O and scale factor $r = 2$. Label the dilated angle, $\angle A'B'C'$.

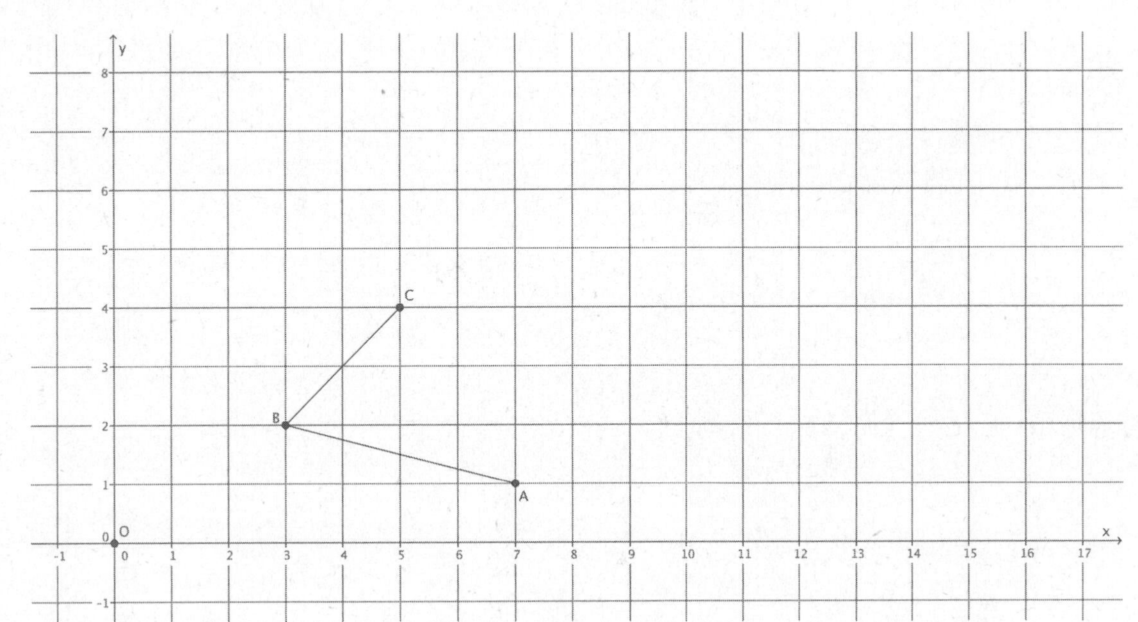

1. If $\angle ABC = 72°$, then what is the measure of $\angle A'B'C'$?

2. If the length of segment AB is 2 cm, what is the length of segment $A'B'$?

3. Which segments, if any, are parallel?

©2019 Great Minds®. eureka-math.org

1. A dilation from center O by scale factor r of an angle maps to what? Verify your claim on the coordinate plane.

The dilation of an angle maps to an angle.

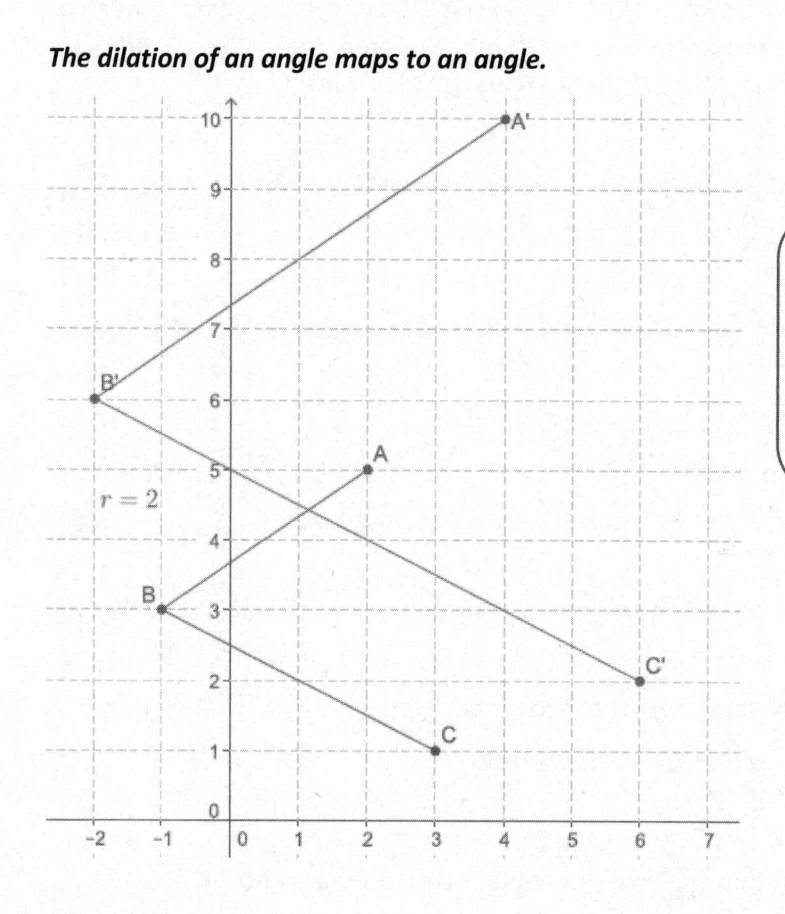

To verify, I can choose any three points to create the angle and any scale factor. I can use what I learned in the last lesson to find the coordinates of the dilated points.

©2019 Great Minds®. eureka-math.org

2. Prove the theorem: A dilation maps rays to rays.

   Let there be a dilation from center O with scale factor r so that $P' = Dilation(P)$ and $Q' = Dilation(Q)$. Show that ray PQ maps to ray $P'Q'$ (i.e., that dilations map rays to rays). Using the diagram, answer the questions that follow, and then informally prove the theorem. (Hint: This proof is a lot like the proof for segments. This time, let U be a point on line PQ that is not between points P and Q.)

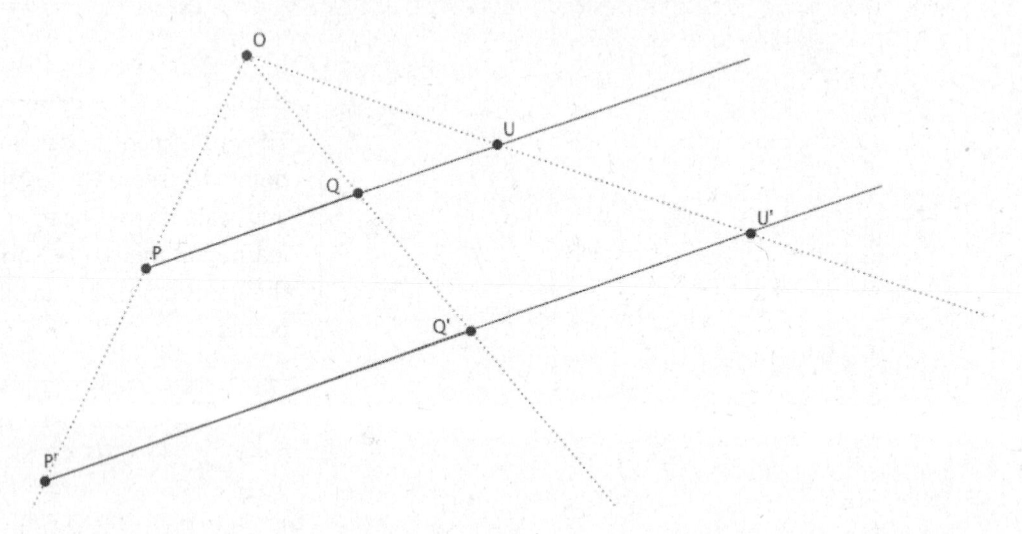

   a. U is a point on $\overrightarrow{PQ}$. By definition of dilation what is the name of $Dilation(U)$?

      By the definition of dilation, we know that $U' = Dilation(U)$.

   b. By the definition of dilation we know that $\frac{|OP'|}{|OP|} = r$. What other two ratios are also equal to r?

      By the definition of dilation, we know that $\frac{|OQ'|}{|OQ|} = \frac{|OU'|}{|OU|} = r$.

   c. By FTS, what do we know about line PQ and line $P'Q'$?

      By FTS, we know that line PQ and line $P'Q'$ are parallel.

   d. What does FTS tell us about line QU and line $Q'U'$?

      By FTS, we know that line QU and line $Q'U'$ are parallel.

   e. What conclusion can be drawn about the line that contains $\overrightarrow{PQ}$ and the line $Q'U'$?

      The line that contains $\overrightarrow{PQ}$ is parallel to $\overleftrightarrow{Q'U'}$ because $\overleftrightarrow{Q'U'}$ is contained in the line $P'Q'$.

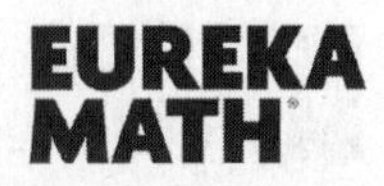

©2019 Great Minds®. eureka-math.org

f. Informally prove that $\overrightarrow{P'Q'}$ is a dilation of $\overrightarrow{PQ}$.

Using the information from parts (a)–(e), we know that U is a point on $\overrightarrow{PQ}$. We also know that the line that contains $\overrightarrow{PQ}$ is parallel to line $Q'U'$. But we already know that $\overleftrightarrow{PQ}$ is parallel to $\overleftrightarrow{P'Q'}$. Since there can only be one line that passes through Q' that is parallel to $\overrightarrow{PQ}$, then the line that contains $\overrightarrow{P'Q'}$ and line $Q'U'$ must coincide. That places the dilation of point U, U', on $\overrightarrow{P'Q'}$, which proves that dilations map rays to rays.

©2019 Great Minds®. eureka-math.org

1. A dilation from center O by scale factor r of a line maps to what? Verify your claim on the coordinate plane.

2. A dilation from center O by scale factor r of a segment maps to what? Verify your claim on the coordinate plane.

3. A dilation from center O by scale factor r of a ray maps to what? Verify your claim on the coordinate plane.

4. Challenge Problem:

   Prove the theorem: *A dilation maps lines to lines.*

   Let there be a dilation from center O with scale factor r so that $P' = Dilation(P)$ and $Q' = Dilation(Q)$. Show that line PQ maps to line $P'Q'$ (i.e., that dilations map lines to lines). Draw a diagram, and then write your informal proof of the theorem. (Hint: This proof is a lot like the proof for segments. This time, let U be a point on line PQ that is not between points P and Q.)

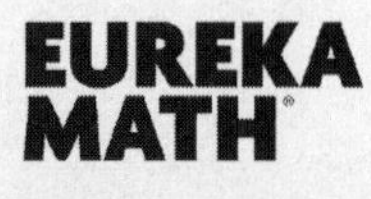

©2019 Great Minds®. eureka-math.org

Example 1

In the picture below, we have a triangle ABC that has been dilated from center O by a scale factor of $r = \frac{1}{2}$. It is noted by $A'B'C'$. We also have triangle $A''B''C''$, which is congruent to triangle $A'B'C'$ (i.e., $\triangle A'B'C' \cong \triangle A''B''C''$).

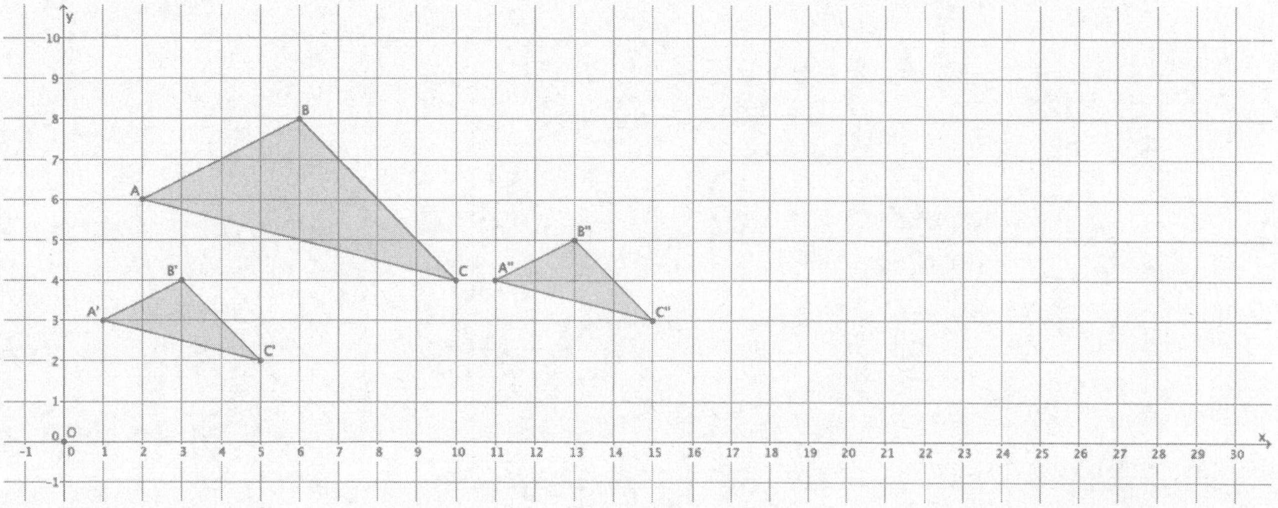

Describe the sequence that would map triangle $A''B''C''$ onto triangle ABC.

©2019 Great Minds®. eureka-math.org

Exercises

1. Triangle ABC was dilated from center O by scale factor $r = \frac{1}{2}$. The dilated triangle is noted by $A'B'C'$. Another triangle $A''B''C''$ is congruent to triangle $A'B'C'$ (i.e., $\triangle A''B''C'' \cong \triangle A'B'C'$). Describe a dilation followed by the basic rigid motion that would map triangle $A''B''C''$ onto triangle ABC.

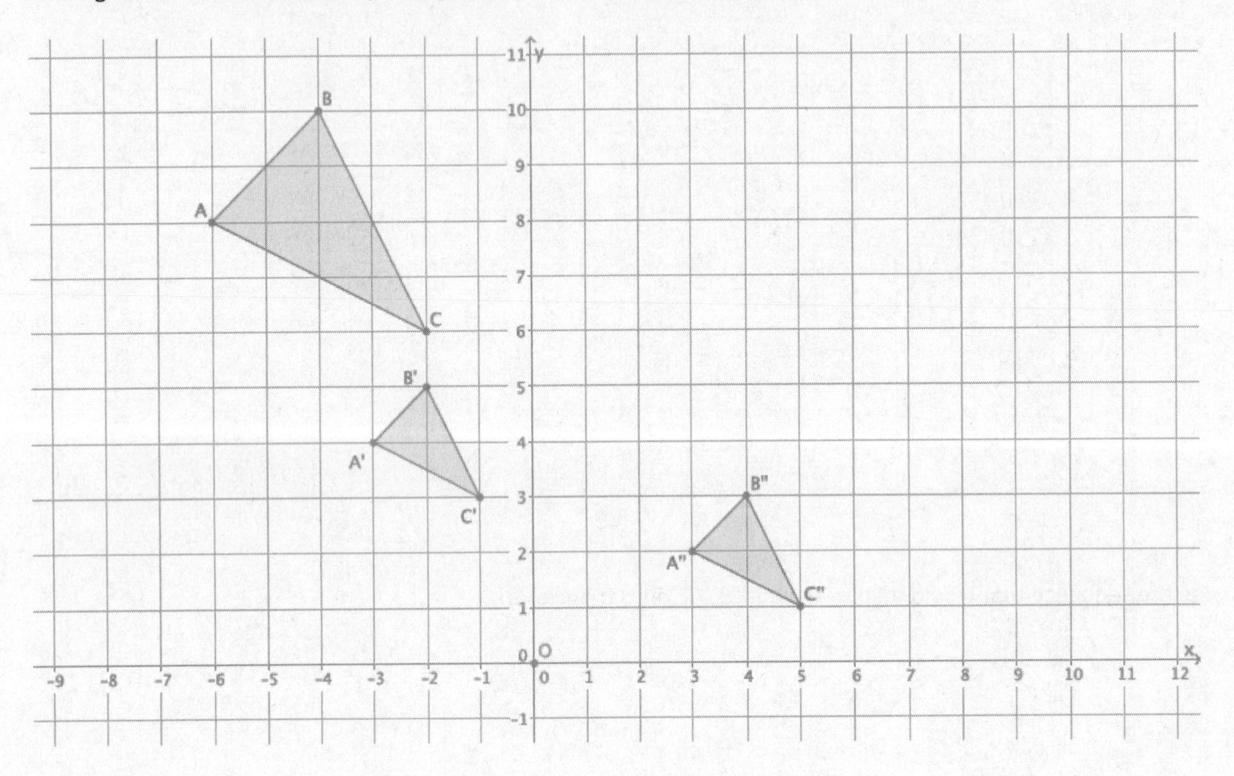

©2019 Great Minds®. eureka-math.org

2. Describe a sequence that would show $\triangle ABC \sim \triangle A'B'C'$.

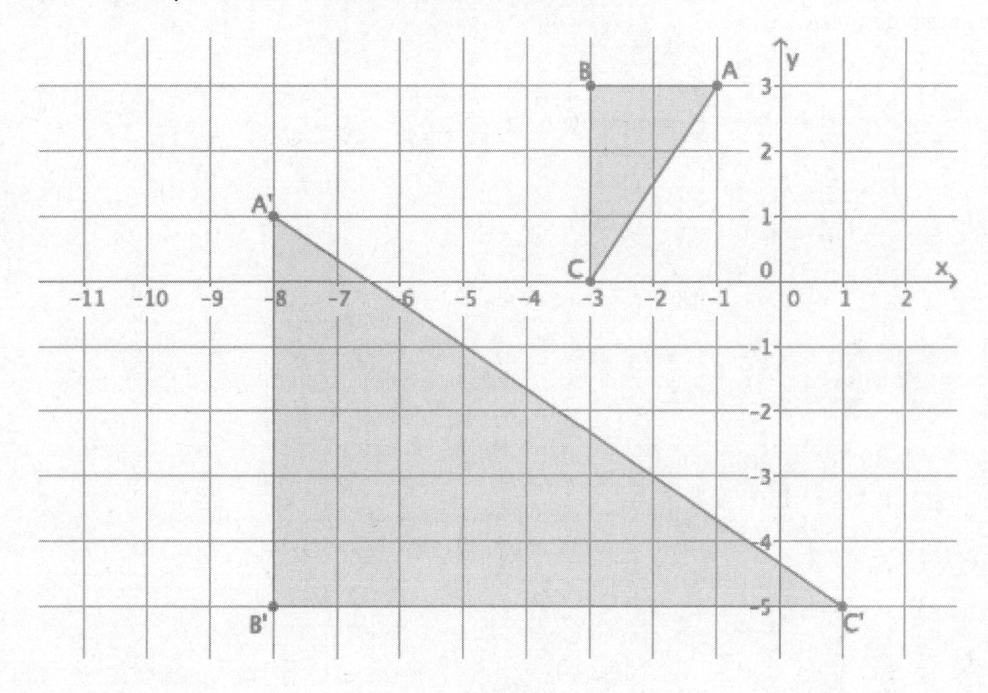

3. Are the two triangles shown below similar? If so, describe a sequence that would prove $\triangle ABC \sim \triangle A'B'C'$. If not, state how you know they are not similar.

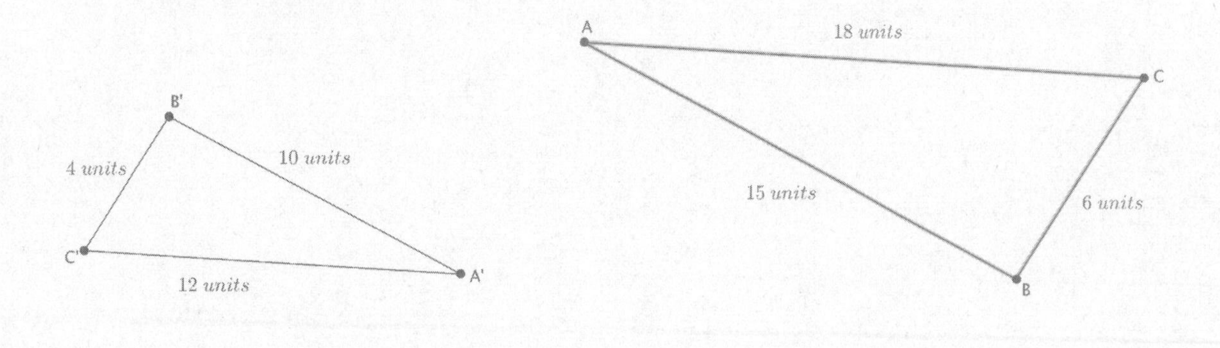

©2019 Great Minds®. eureka-math.org

4. Are the two triangles shown below similar? If so, describe a sequence that would prove $\triangle ABC \sim \triangle A'B'C'$. If not, state how you know they are not similar.

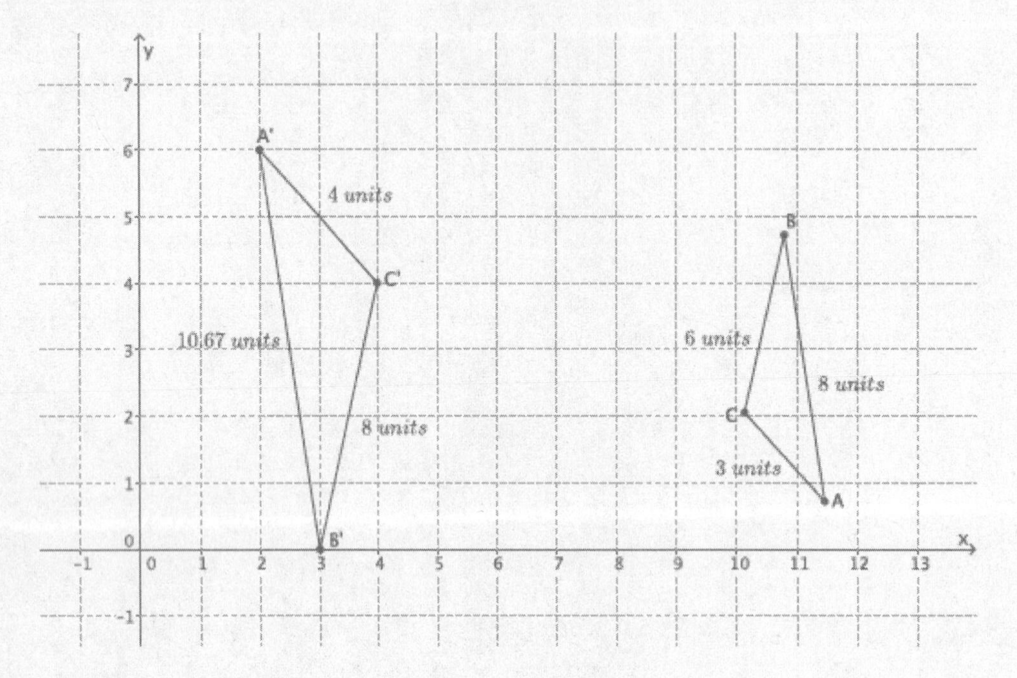

©2019 Great Minds®. eureka-math.org

Lesson Summary

A *similarity transformation* (or a *similarity*) is a sequence of a finite number of dilations or basic rigid motions. Two figures are *similar* if there is a similarity transformation taking one figure onto the other figure. Every similarity can be represented as a dilation followed by a congruence.

The notation $\triangle ABC \sim \triangle A'B'C'$ means that $\triangle ABC$ is similar to $\triangle A'B'C'$.

©2019 Great Minds®. eureka-math.org

Name ____________________ Date ____________

In the picture below, we have triangle DEF that has been dilated from center O by scale factor $r = \frac{1}{2}$. The dilated triangle is noted by $D'E'F'$. We also have a triangle $D''EF$, which is congruent to triangle DEF (i.e., $\triangle DEF \cong \triangle D''EF$). Describe the sequence of a dilation followed by a congruence (of one or more rigid motions) that would map triangle $D'E'F'$ onto triangle $D''EF$.

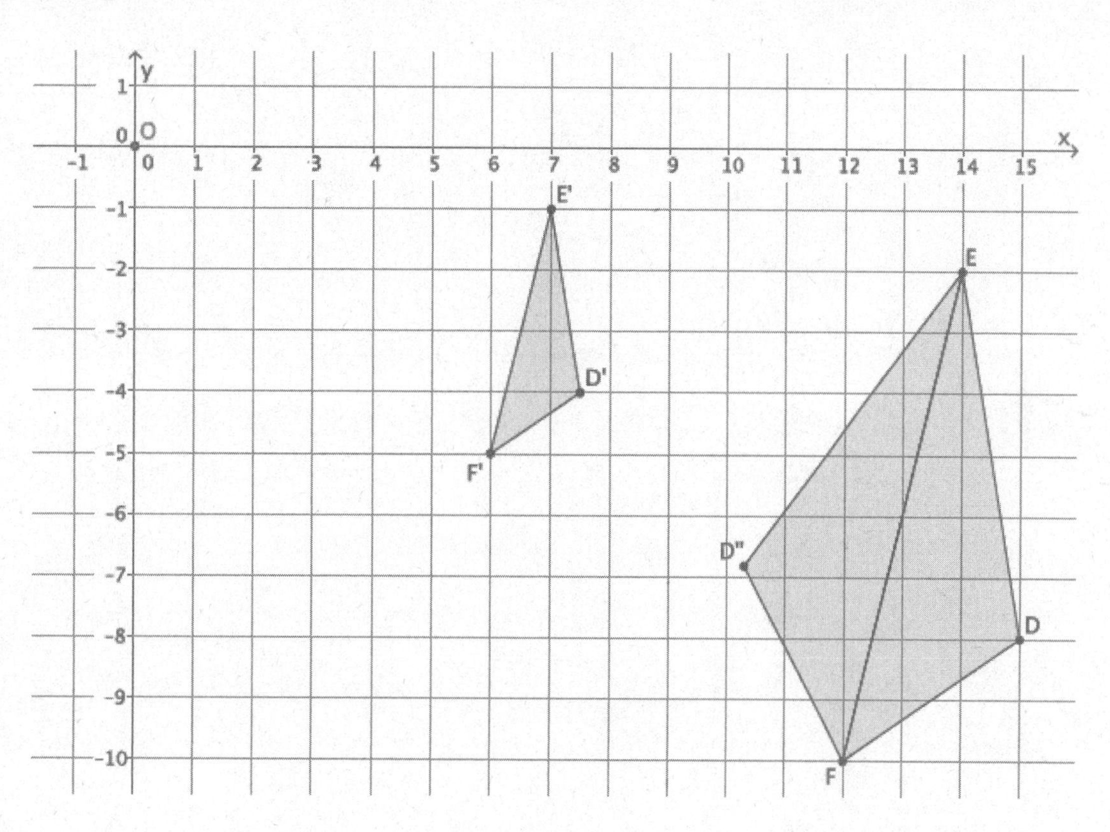

©2019 Great Minds®. eureka-math.org

1. In the picture below, we have a triangle ABC that has been dilated from center O by scale factor $r = 3$. It is noted by $A'B'C'$. We also have a triangle $A''B''C''$, which is congruent to triangle $A'B'C'$ (i.e., $\triangle A'B'C' \cong \triangle A''B''C''$). Describe the sequence of a dilation, followed by a congruence (of one or more rigid motions), that would map triangle $A''B''C''$ onto triangle ABC.

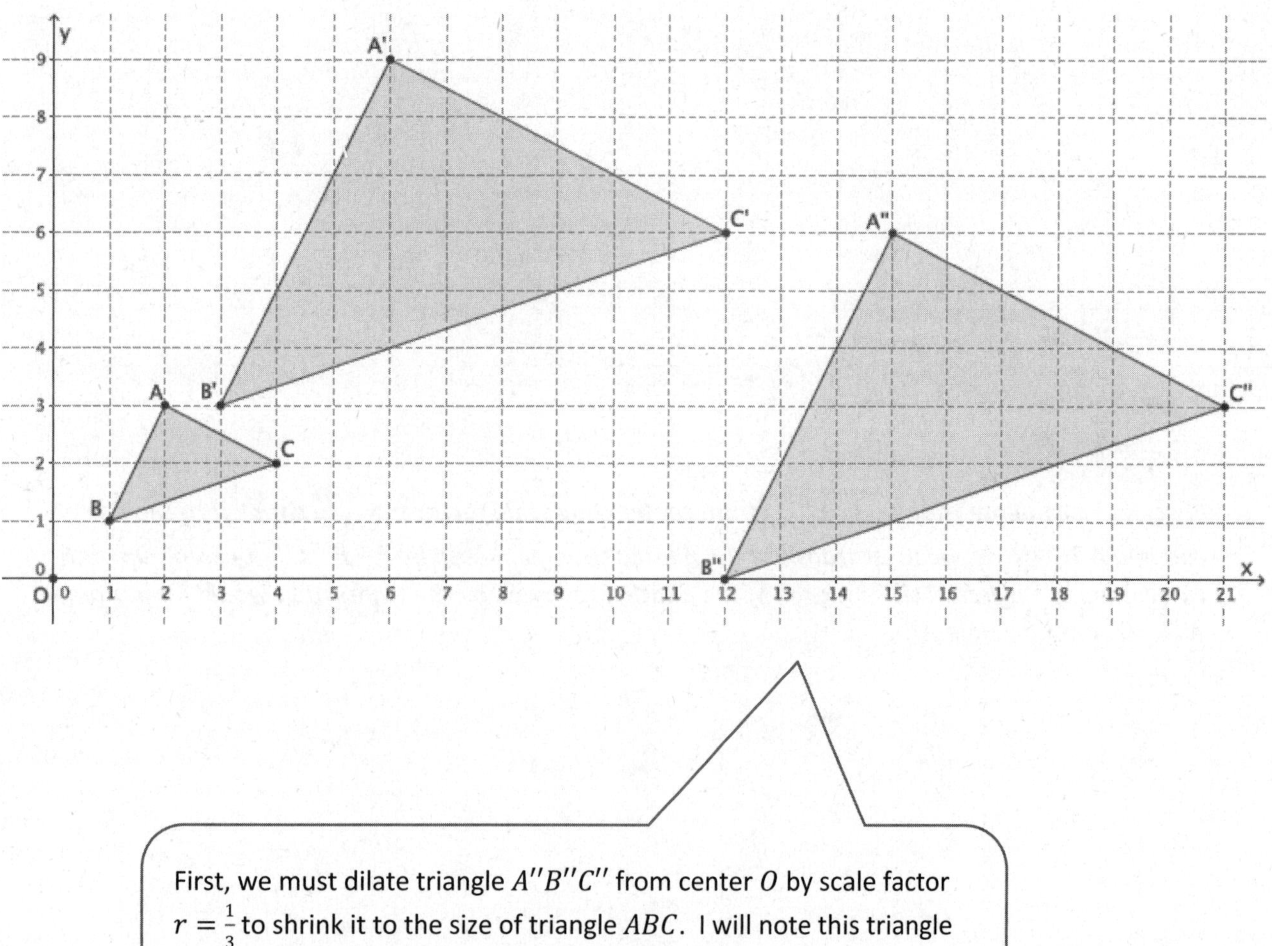

First, we must dilate triangle $A''B''C''$ from center O by scale factor $r = \frac{1}{3}$ to shrink it to the size of triangle ABC. I will note this triangle as $A'''B'''C'''$. Once I have the triangle to the right size, I can translate the dilated triangle, $A'''B'''C'''$, up one unit and to the left three units.

©2019 Great Minds®. eureka-math.org

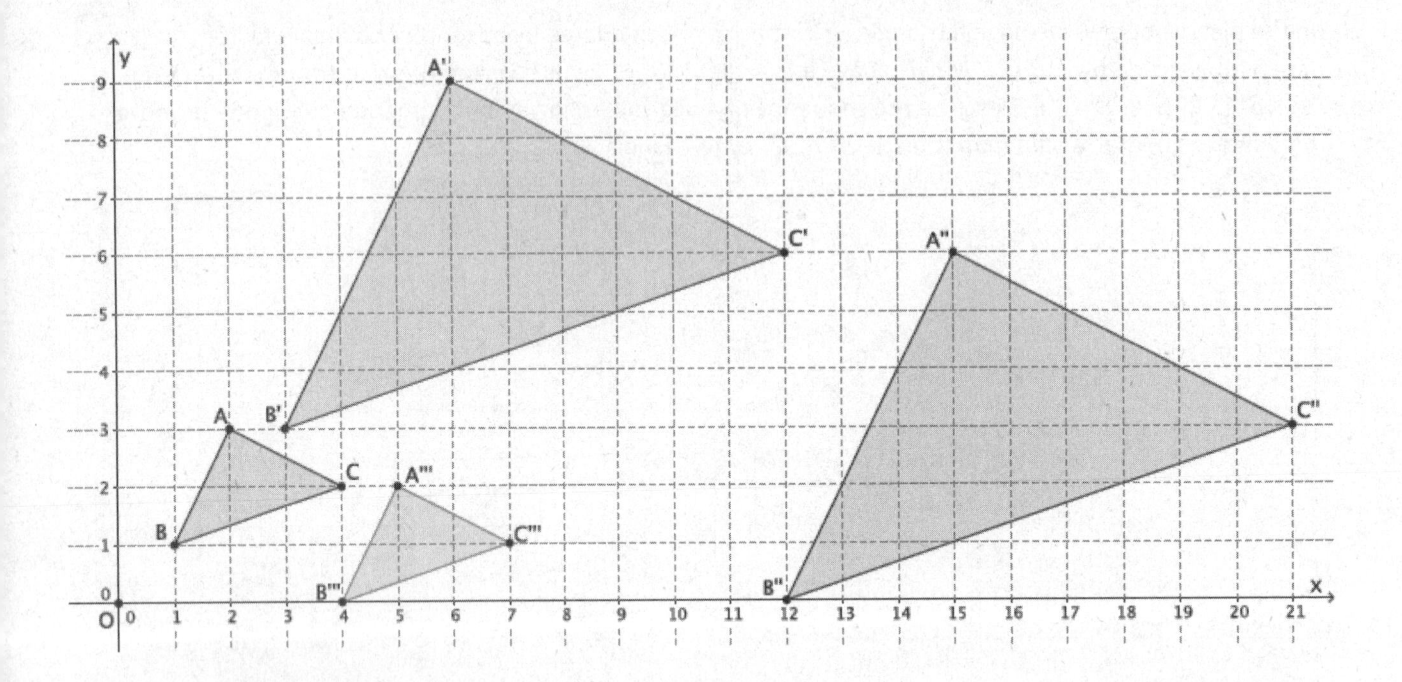

First, we must dilate triangle $A''B''C''$ from center O by scale factor $r = \frac{1}{3}$ to shrink it to the size of triangle ABC. Next, we must translate the dilated triangle, noted by $A'''B'''C'''$, one unit up and three units to the left. This sequence of the dilation followed by the translation would map triangle $A''B''C''$ onto triangle ABC.

©2019 Great Minds®. eureka-math.org

2. Triangle ABC is similar to triangle $A'B'C'$ (i.e., $\triangle ABC \sim \triangle A'B'C'$). Prove the similarity by describing a sequence that would map triangle $A'B'C'$ onto triangle ABC.

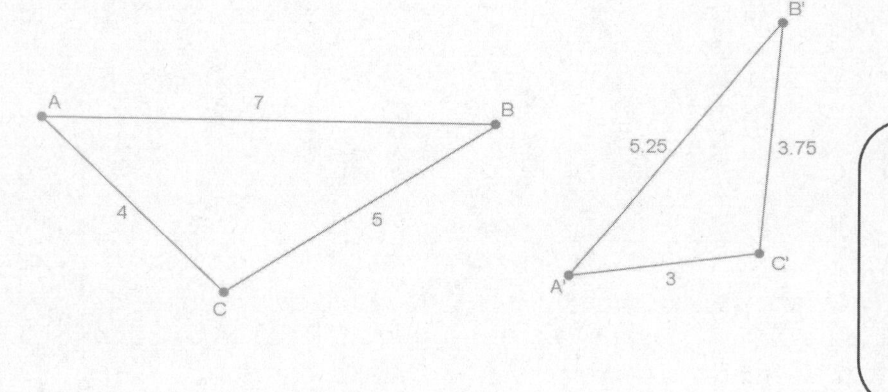

I can check the ratios of the corresponding sides to see if they are the same proportion and equal to the same scale factor.

The scale factor that would magnify triangle $A'B'C'$ to the size of triangle ABC is $r = \frac{4}{3}$.

Once the triangle $A'B'C'$ is the same size as triangle ABC, I can describe a congruence to map triangle $A'B'C'$ onto triangle ABC.

Sample description:

The sequence that would prove the similarity of the triangles is a dilation from a center by a scale factor of $r = \frac{4}{3}$, followed by a translation along vector $\overrightarrow{A'A}$, and finally, a rotation about point A.

©2019 Great Minds®. eureka-math.org

1. In the picture below, we have triangle DEF that has been dilated from center O by scale factor $r = 4$. It is noted by $D'E'F'$. We also have triangle $D''E''F''$, which is congruent to triangle $D'E'F'$ (i.e., $\triangle D'E'F' \cong \triangle D''E''F''$). Describe the sequence of a dilation, followed by a congruence (of one or more rigid motions), that would map triangle $D''E''F''$ onto triangle DEF.

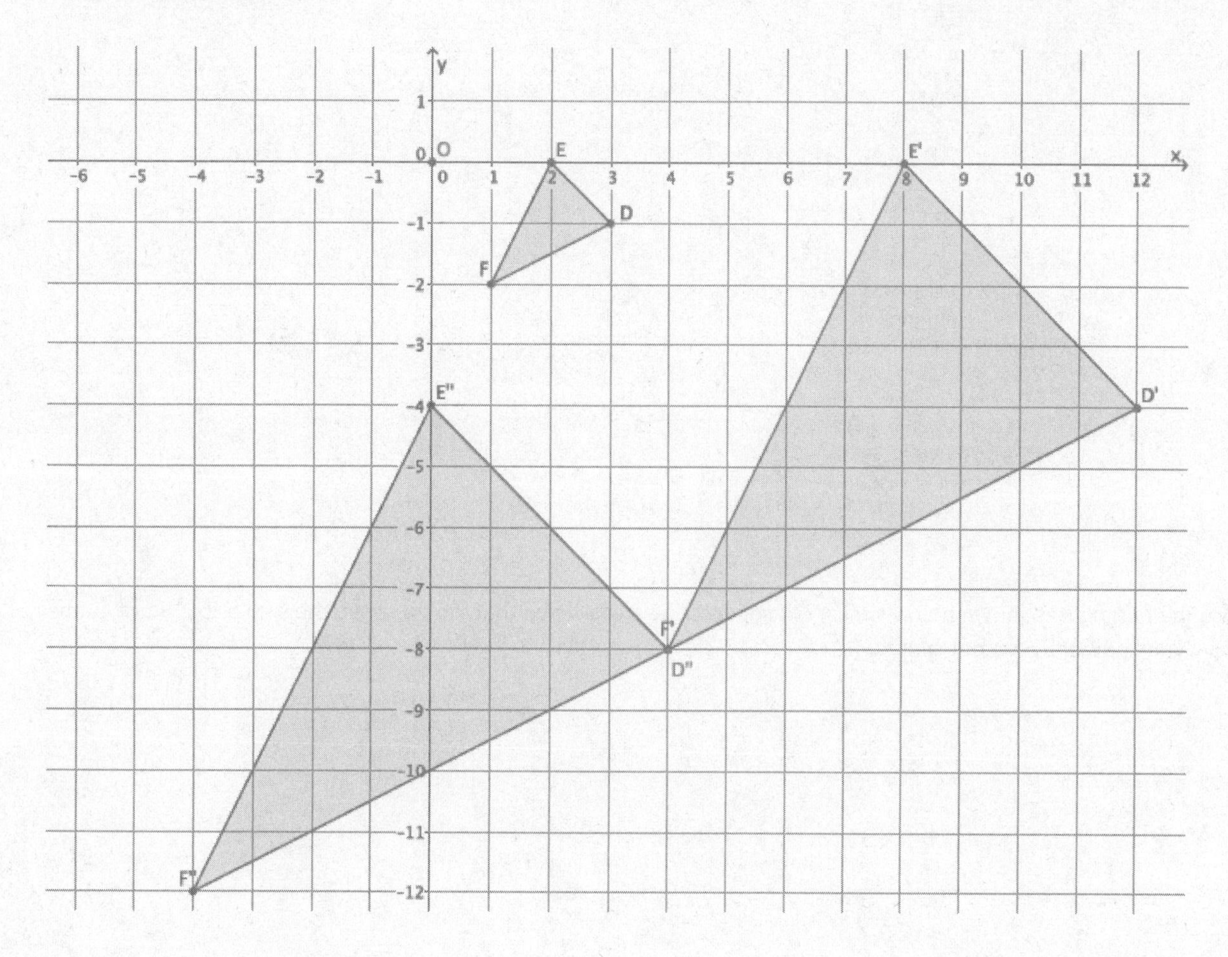

©2019 Great Minds®. eureka-math.org

2. Triangle ABC was dilated from center O by scale factor $r = \frac{1}{2}$. The dilated triangle is noted by $A'B'C'$. Another triangle $A''B''C''$ is congruent to triangle $A'B'C'$ (i.e., $\triangle A''B''C'' \cong \triangle A'B'C'$). Describe the dilation followed by the basic rigid motions that would map triangle $A''B''C''$ onto triangle ABC.

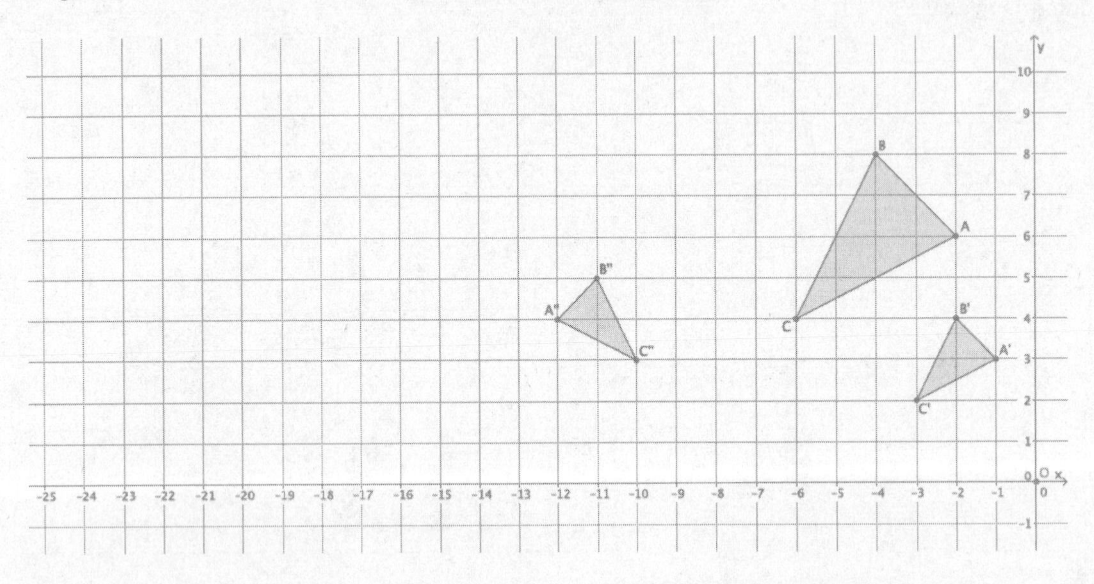

3. Are the two figures shown below similar? If so, describe a sequence that would prove the similarity. If not, state how you know they are not similar.

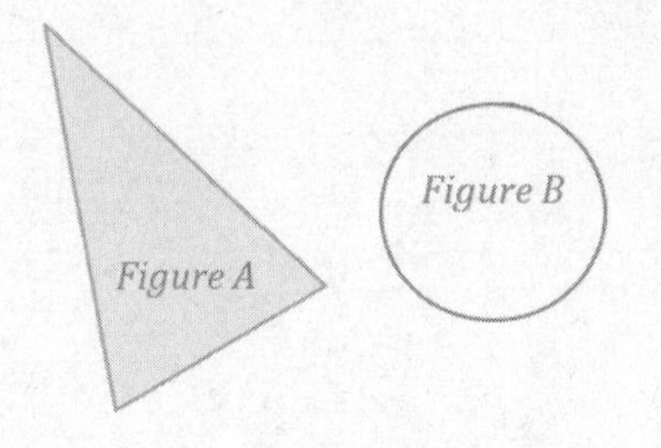

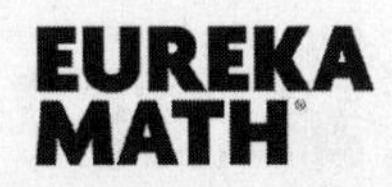

©2019 Great Minds®. eureka-math.org

4. Triangle ABC is similar to triangle $A'B'C'$ (i.e., $\triangle ABC \sim \triangle A'B'C'$). Prove the similarity by describing a sequence that would map triangle $A'B'C'$ onto triangle ABC.

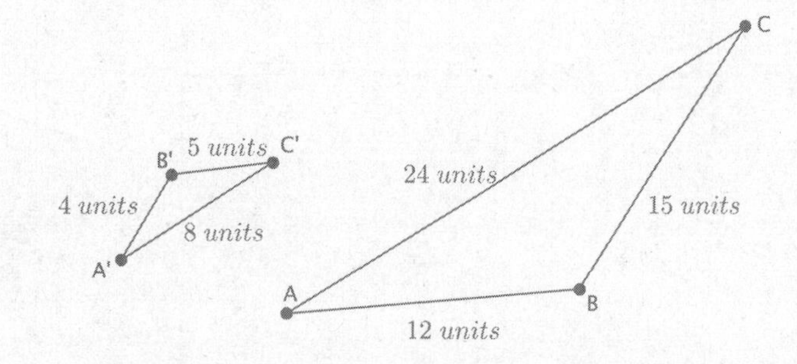

5. Are the two figures shown below similar? If so, describe a sequence that would prove $\triangle ABC \sim \triangle A'B'C'$. If not, state how you know they are not similar.

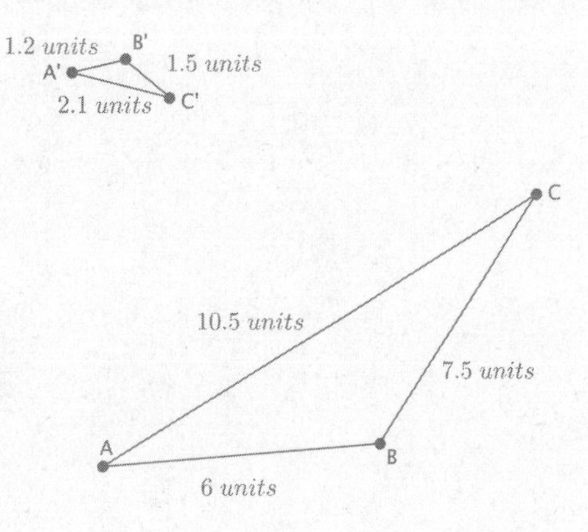

©2019 Great Minds®. eureka-math.org

6. Describe a sequence that would show $\triangle ABC \sim \triangle A'B'C'$.

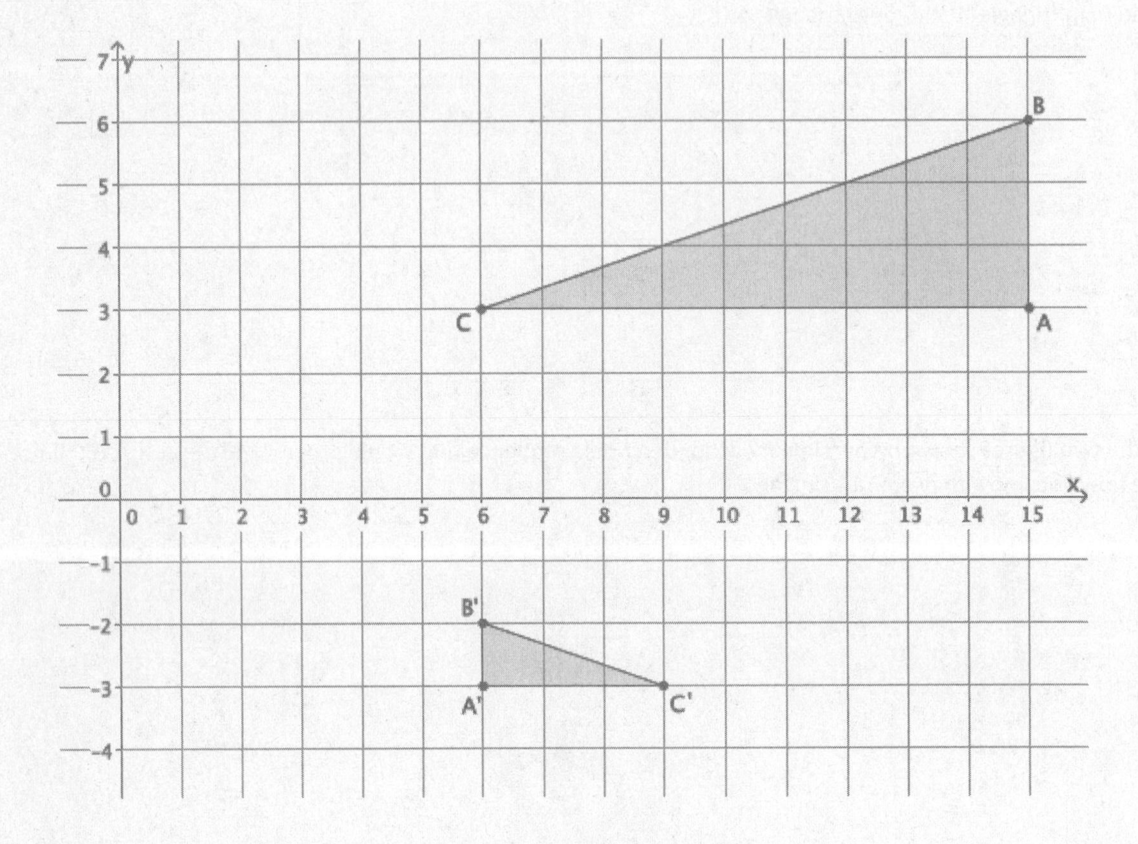

©2019 Great Minds®. eureka-math.org

Exploratory Challenge 1

The goal is to show that if $\triangle ABC$ is similar to $\triangle A'B'C'$, then $\triangle A'B'C'$ is similar to $\triangle ABC$. Symbolically, if $\triangle ABC \sim \triangle A'B'C'$, then $\triangle A'B'C' \sim \triangle ABC$.

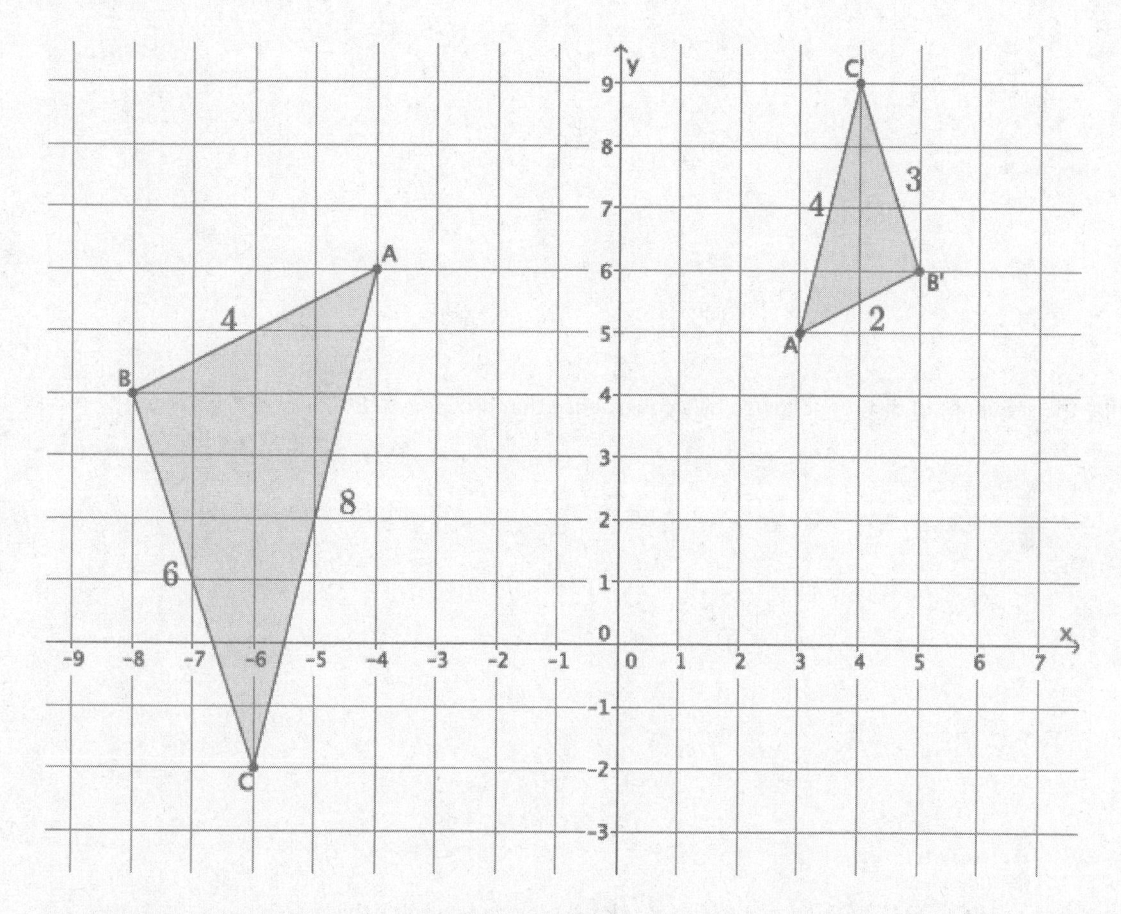

a. First, determine whether or not $\triangle ABC$ is in fact similar to $\triangle A'B'C'$. (If it isn't, then no further work needs to be done.) Use a protractor to verify that the corresponding angles are congruent and that the ratios of the corresponding sides are equal to some scale factor.

©2019 Great Minds®. eureka-math.org

b. Describe the sequence of dilation followed by a congruence that proves $\triangle ABC \sim \triangle A'B'C'$.

c. Describe the sequence of dilation followed by a congruence that proves $\triangle A'B'C' \sim \triangle ABC$.

d. Is it true that $\triangle ABC \sim \triangle A'B'C'$ and $\triangle A'B'C' \sim \triangle ABC$? Why do you think this is so?

©2019 Great Minds®. eureka-math.org

Exploratory Challenge 2

The goal is to show that if $\triangle ABC$ is similar to $\triangle A'B'C'$ and $\triangle A'B'C'$ is similar to $\triangle A''B''C''$, then $\triangle ABC$ is similar to $\triangle A''B''C''$. Symbolically, if $\triangle ABC \sim \triangle A'B'C'$ and $\triangle A'B'C' \sim \triangle A''B''C''$, then $\triangle ABC \sim \triangle A''B''C''$.

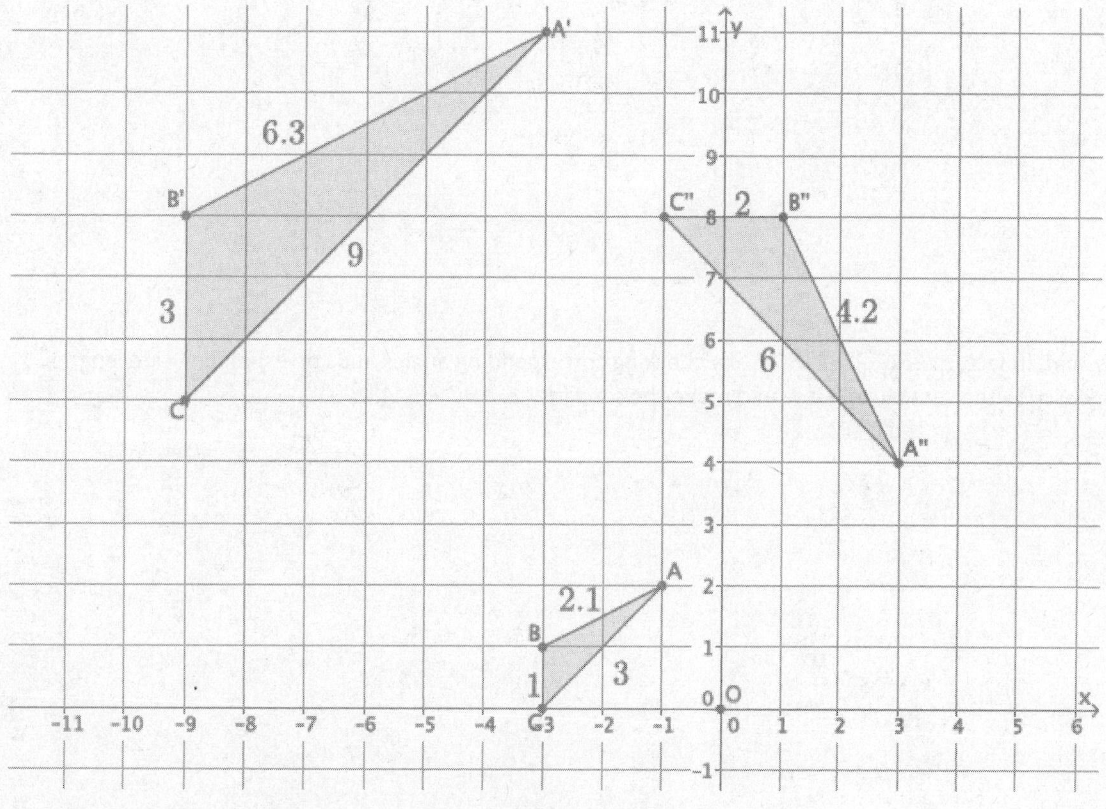

a. Describe the similarity that proves $\triangle ABC \sim \triangle A'B'C'$.

©2019 Great Minds®. eureka-math.org

b. Describe the similarity that proves $\triangle A'B'C' \sim \triangle A''B''C''$.

c. Verify that, in fact, $\triangle ABC \sim \triangle A''B''C''$ by checking corresponding angles and corresponding side lengths. Then, describe the sequence that would prove the similarity $\triangle ABC \sim \triangle A''B''C''$.

d. Is it true that if $\triangle ABC \sim \triangle A'B'C'$ and $\triangle A'B'C' \sim \triangle A''B''C''$, then $\triangle ABC \sim \triangle A''B''C''$? Why do you think this is so?

©2019 Great Minds®. eureka-math.org

Lesson Summary

Similarity is a symmetric relation. That means that if one figure is similar to another, $S \sim S'$, then we can be sure that $S' \sim S$.

Similarity is a transitive relation. That means that if we are given two similar figures, $S \sim T$, and another statement about $T \sim U$, then we also know that $S \sim U$.

©2019 Great Minds®. eureka-math.org

Name ______________________________ Date ______________

Use the diagram below to answer Problems 1 and 2.

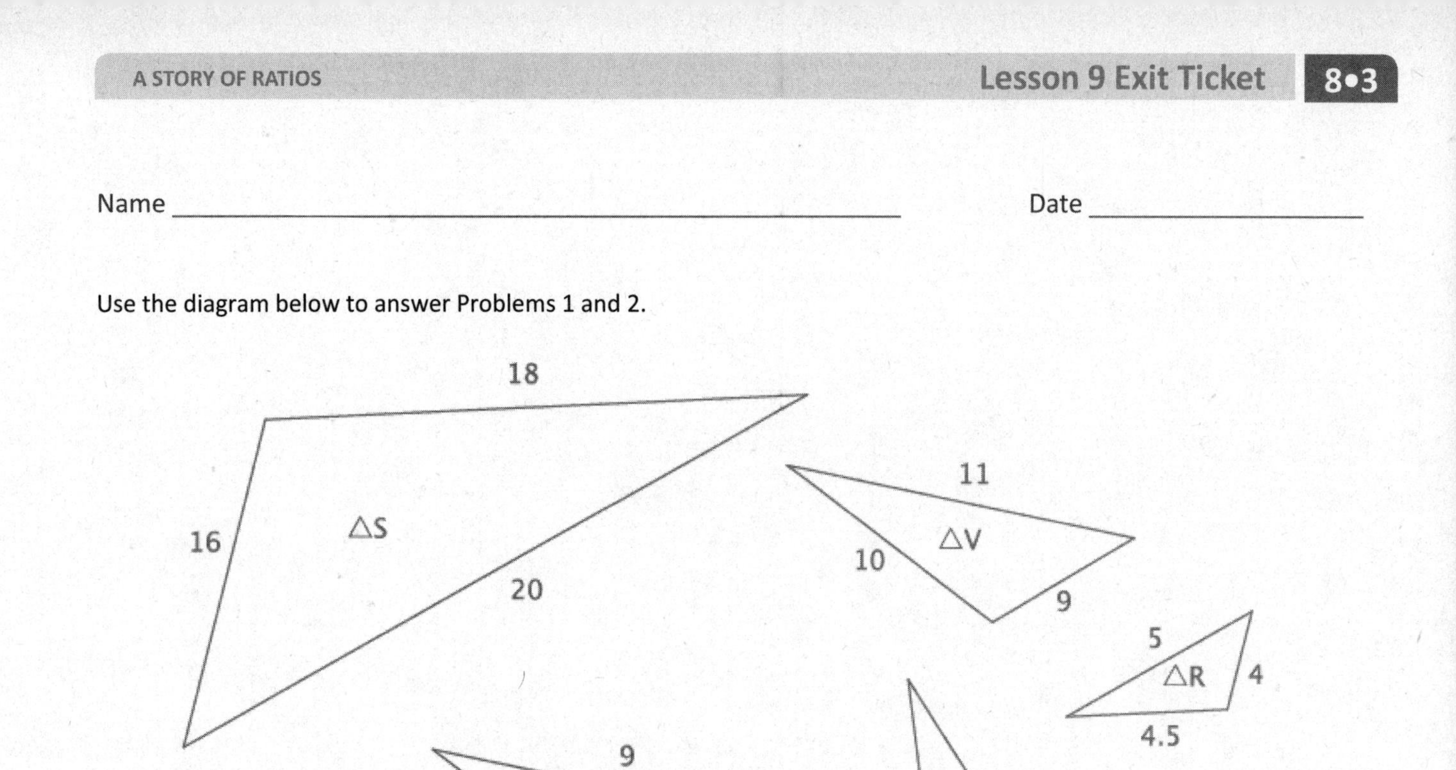

1. Which two triangles, if any, have similarity that is symmetric?

2. Which three triangles, if any, have similarity that is transitive?

©2019 Great Minds®. eureka-math.org

1. In the diagram below, $\triangle ABC \sim \triangle A'B'C'$ and $\triangle A'B'C' \sim \triangle A''B''C''$. Is $\triangle ABC \sim \triangle A''B''C''$? If so, describe the dilation followed by the congruence that demonstrates the similarity.

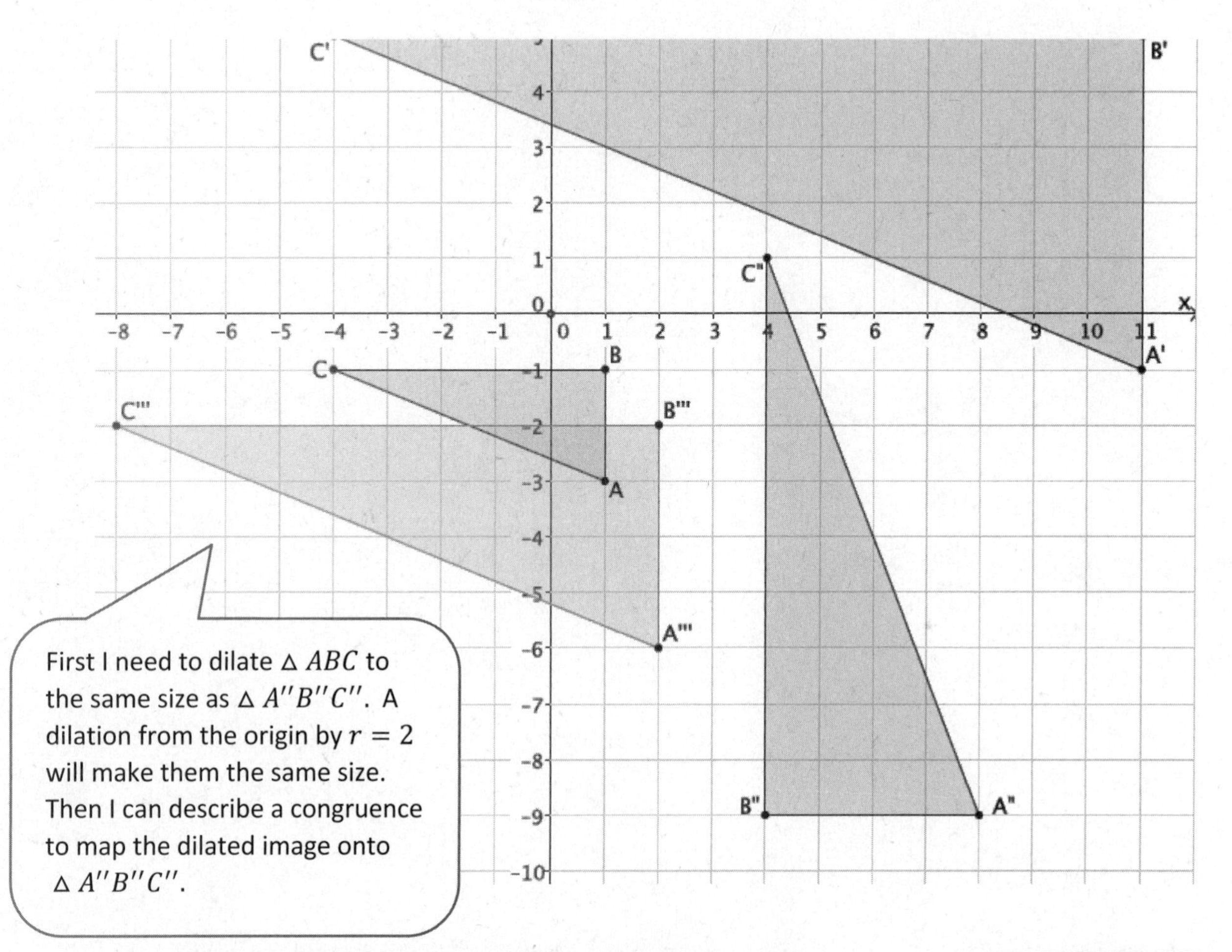

Yes, $\triangle ABC \sim \triangle A''B''C''$ because similarity is transitive. Since $r|AB| = |A''B''|$, $|AB| = 2$, and $|A''B''| = 4$, then $r \cdot 2 = 4$. Therefore, $r = 2$. Then, a dilation from the origin by scale factor $r = 2$ makes $\triangle ABC$ the same size as $\triangle A''B''C''$. Translate the dilated image of $\triangle ABC$, $\triangle A'''B'''C'''$, 12 units to the right and 3 units up to map C''' to C''. Next, rotate the dilated image about point C'', 90 degrees clockwise. Finally, reflect the rotated image across line $C''B''$. The sequence of the dilation and the congruence map $\triangle ABC$ onto $\triangle A''B''C''$, demonstrating the similarity.

I remember that when I have to translate an image, it is best to do it so that corresponding points, like C and C'', coincide.

©2019 Great Minds®. eureka-math.org

1. Would a dilation alone be enough to show that similarity is symmetric? That is, would a dilation alone prove that if $\triangle ABC \sim \triangle A'B'C'$, then $\triangle A'B'C' \sim \triangle ABC$? Consider the two examples below.

   a. Given $\triangle ABC \sim \triangle A'B'C'$, is a dilation enough to show that $\triangle A'B'C' \sim \triangle ABC$? Explain.

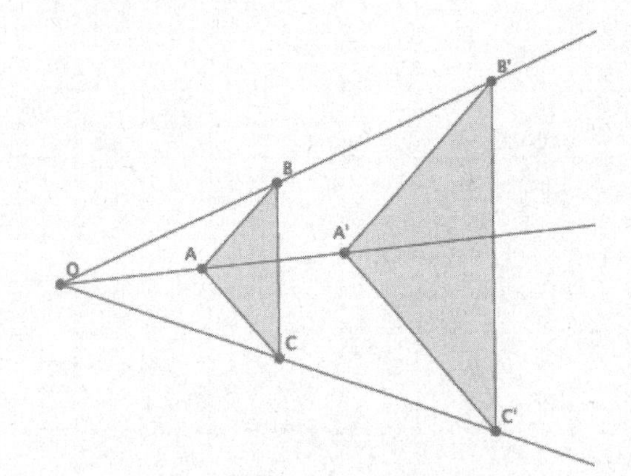

   b. Given $\triangle ABC \sim \triangle A'B'C'$, is a dilation enough to show that $\triangle A'B'C' \sim \triangle ABC$? Explain.

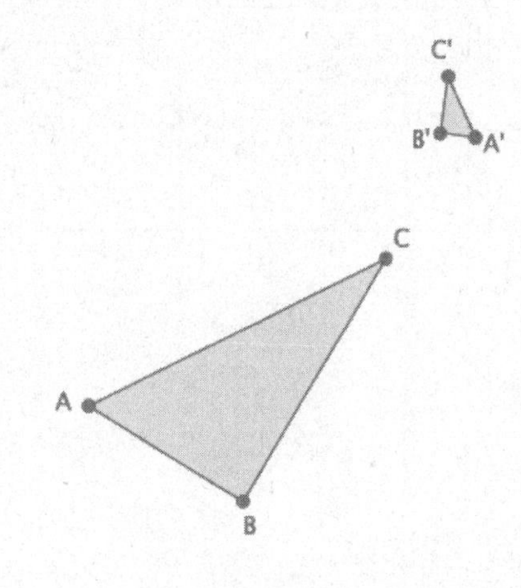

   c. In general, is dilation enough to prove that similarity is a symmetric relation? Explain.

©2019 Great Minds®. eureka-math.org

2. Would a dilation alone be enough to show that similarity is transitive? That is, would a dilation alone prove that if $\triangle ABC \sim \triangle A'B'C'$ and $\triangle A'B'C' \sim \triangle A''B''C''$, then $\triangle ABC \sim \triangle A''B''C''$? Consider the two examples below.

   a. Given $\triangle ABC \sim \triangle A'B'C'$ and $\triangle A'B'C' \sim \triangle A''B''C''$, is a dilation enough to show that $\triangle ABC \sim \triangle A''B''C''$? Explain.

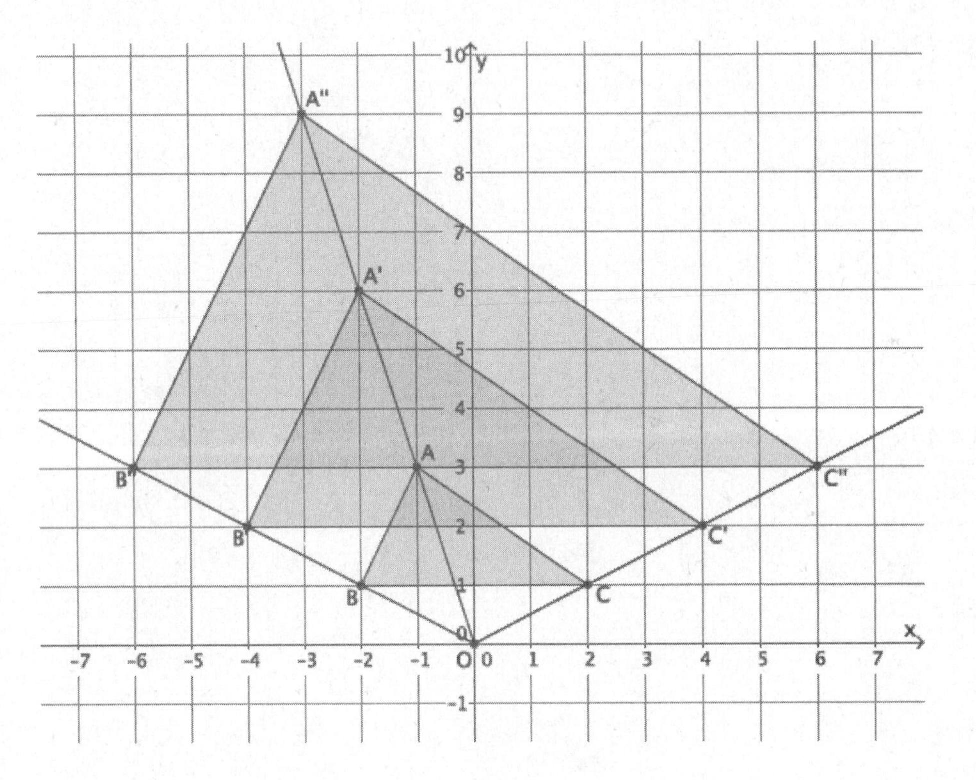

©2019 Great Minds®. eureka-math.org

b. Given $\triangle ABC \sim \triangle A'B'C'$ and $\triangle A'B'C' \sim \triangle A''B''C''$, is a dilation enough to show that $\triangle ABC \sim \triangle A''B''C''$? Explain.

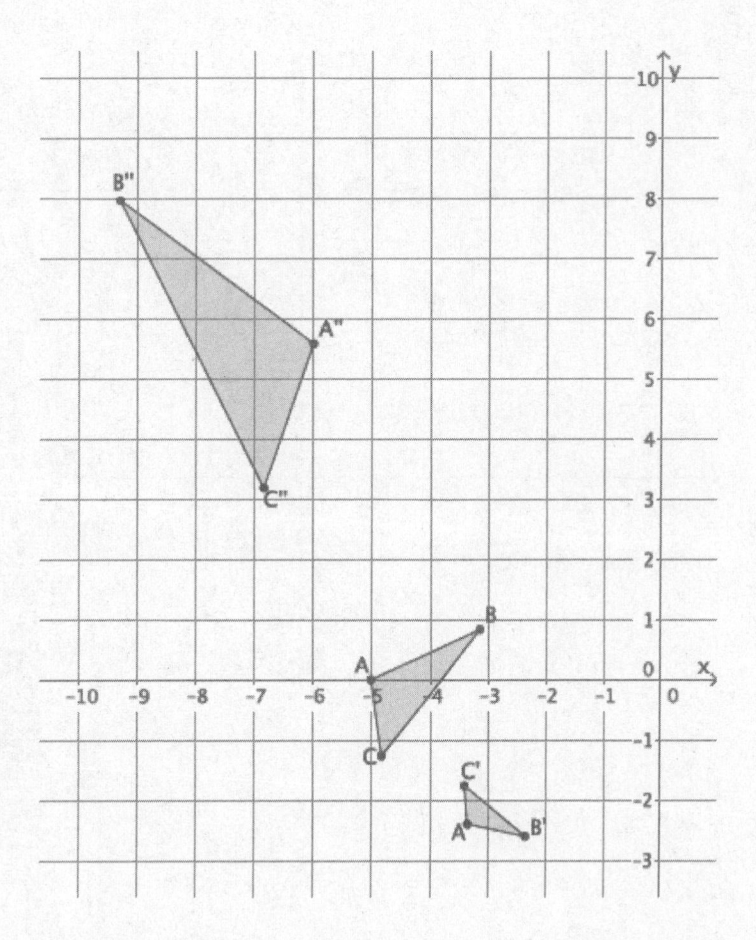

c. In general, is dilation enough to prove that similarity is a transitive relation? Explain.

©2019 Great Minds®. eureka-math.org

3. In the diagram below, $\triangle ABC \sim \triangle A'B'C'$ and $\triangle A'B'C' \sim \triangle A''B''C''$. Is $\triangle ABC \sim \triangle A''B''C''$? If so, describe the dilation followed by the congruence that demonstrates the similarity.

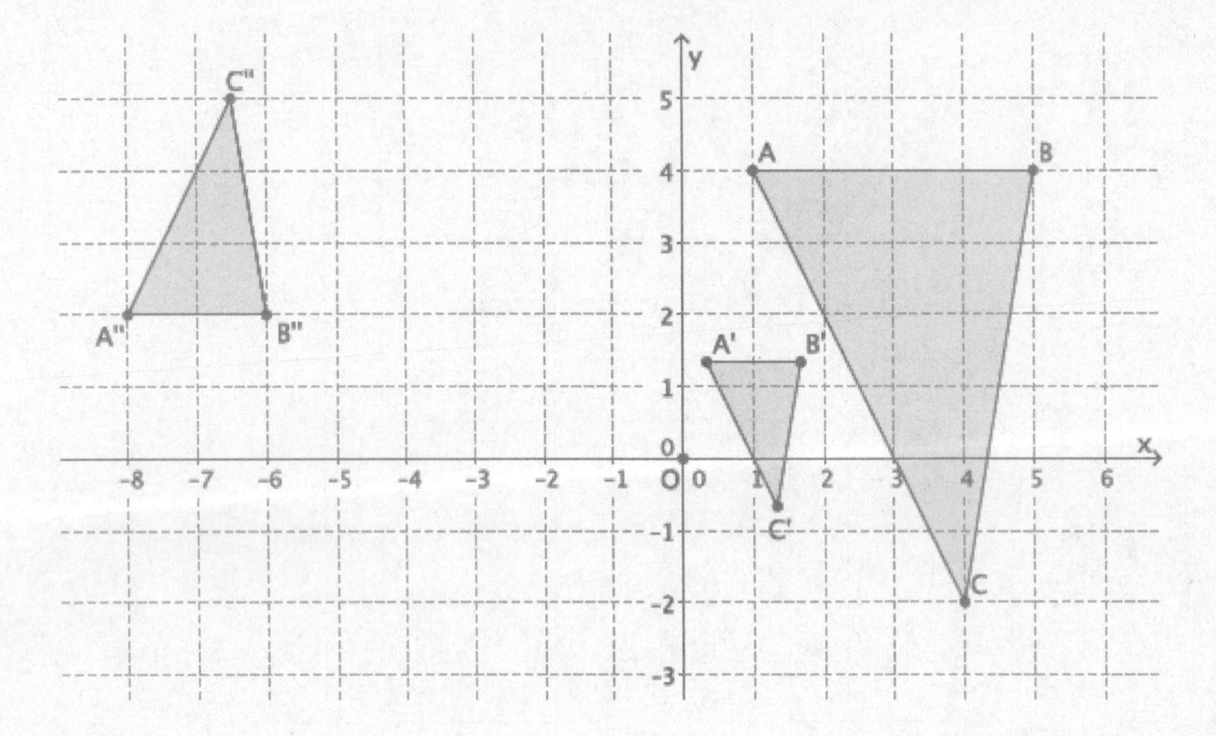

©2019 Great Minds®. eureka-math.org

Exercises 1–5

1. Use a protractor to draw a pair of triangles with two pairs of interior angles that are equal in measure. Then, measure the lengths of the sides, and verify that the lengths of their corresponding sides are equal in ratio.

2. Draw a new pair of triangles with two pairs of interior angles that are equal in measure. Then, measure the lengths of the sides, and verify that the lengths of their corresponding sides are equal in ratio.

©2019 Great Minds®. eureka-math.org

3. Are the triangles shown below similar? Present an informal argument as to why they are or are not similar.

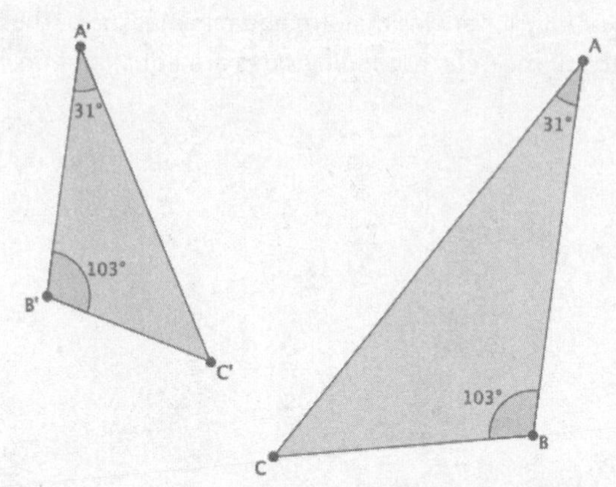

4. Are the triangles shown below similar? Present an informal argument as to why they are or are not similar.

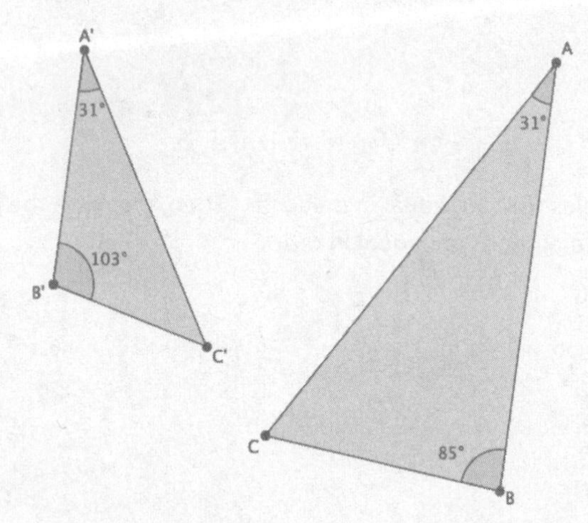

5. Are the triangles shown below similar? Present an informal argument as to why they are or are not similar.

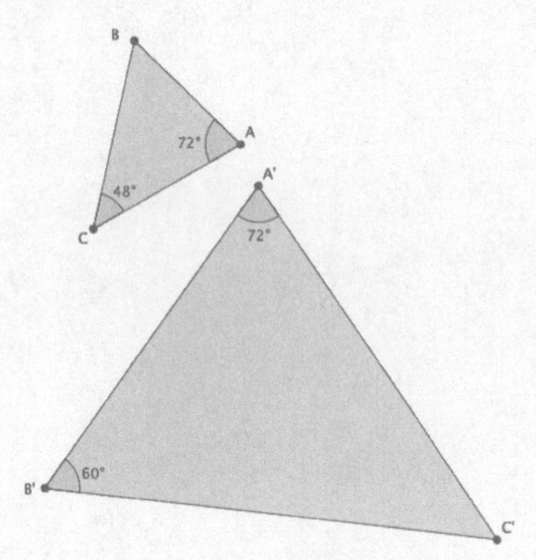

©2019 Great Minds®. eureka-math.org

Lesson Summary

Two triangles are said to be similar if they have two pairs of corresponding angles that are equal in measure.

©2019 Great Minds®. eureka-math.org

Name ____ Date ________________

1. Are the triangles shown below similar? Present an informal argument as to why they are or are not similar.

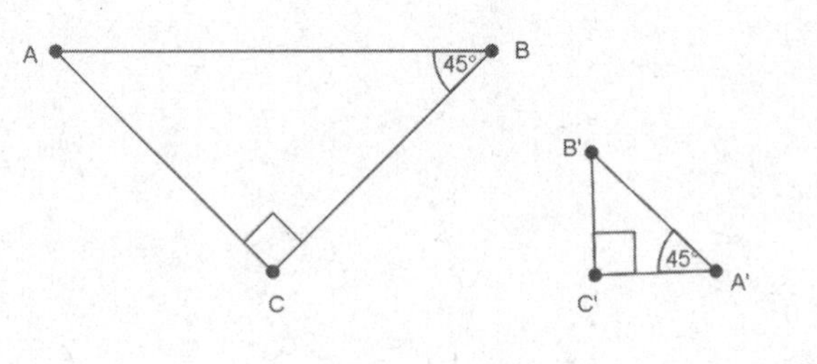

2. Are the triangles shown below similar? Present an informal argument as to why they are or are not similar.

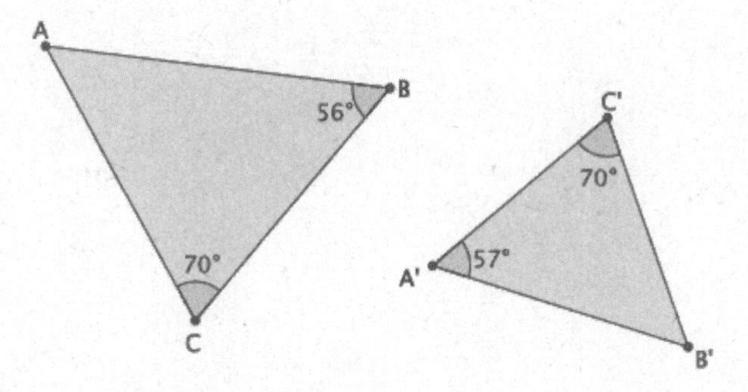

©2019 Great Minds®. eureka-math.org

1. Are the triangles shown below similar? Present an informal argument as to why they are or are not similar.

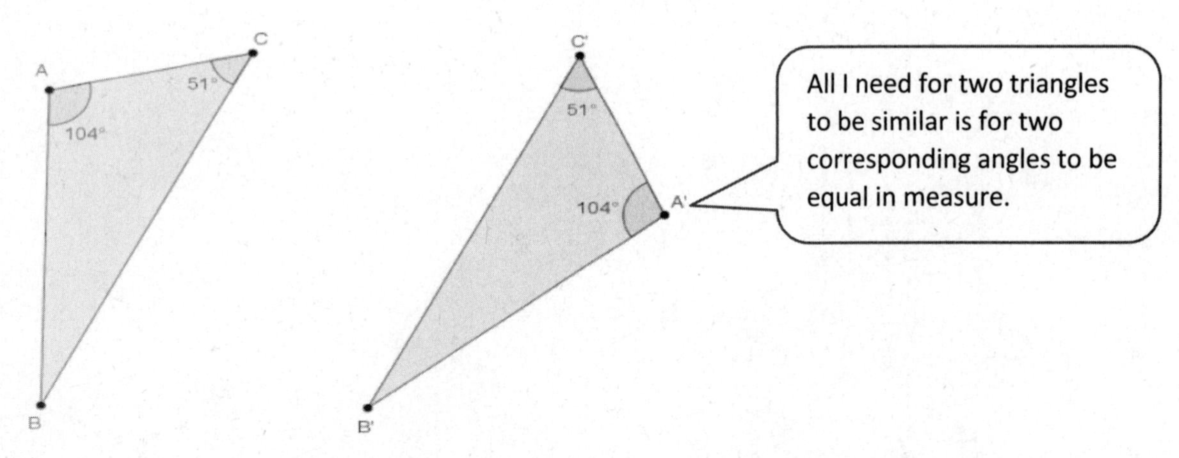

Yes, $\triangle ABC \sim \triangle A'B'C'$. They are similar because they have two pairs of corresponding angles that are equal in measure, namely, $|\angle A| = |\angle A'| = 104°$, and $|\angle C| = |\angle C'| = 51°$.

2. Are the triangles shown below similar? Present an informal argument as to why they are or are not similar.

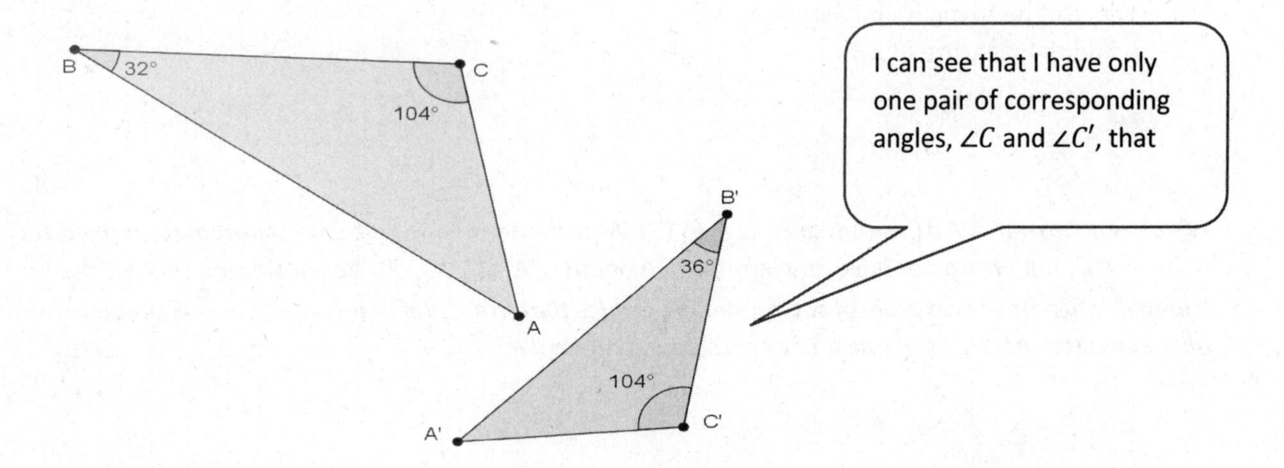

No, $\triangle ABC$ is not similar to $\triangle A'B'C'$. By the given information, $|\angle B| \neq |\angle B'|$, and $|\angle A| \neq |\angle A'|$.

©2019 Great Minds®. eureka-math.org

3. Are the triangles shown below similar? Present an informal argument as to why they are or are not similar.

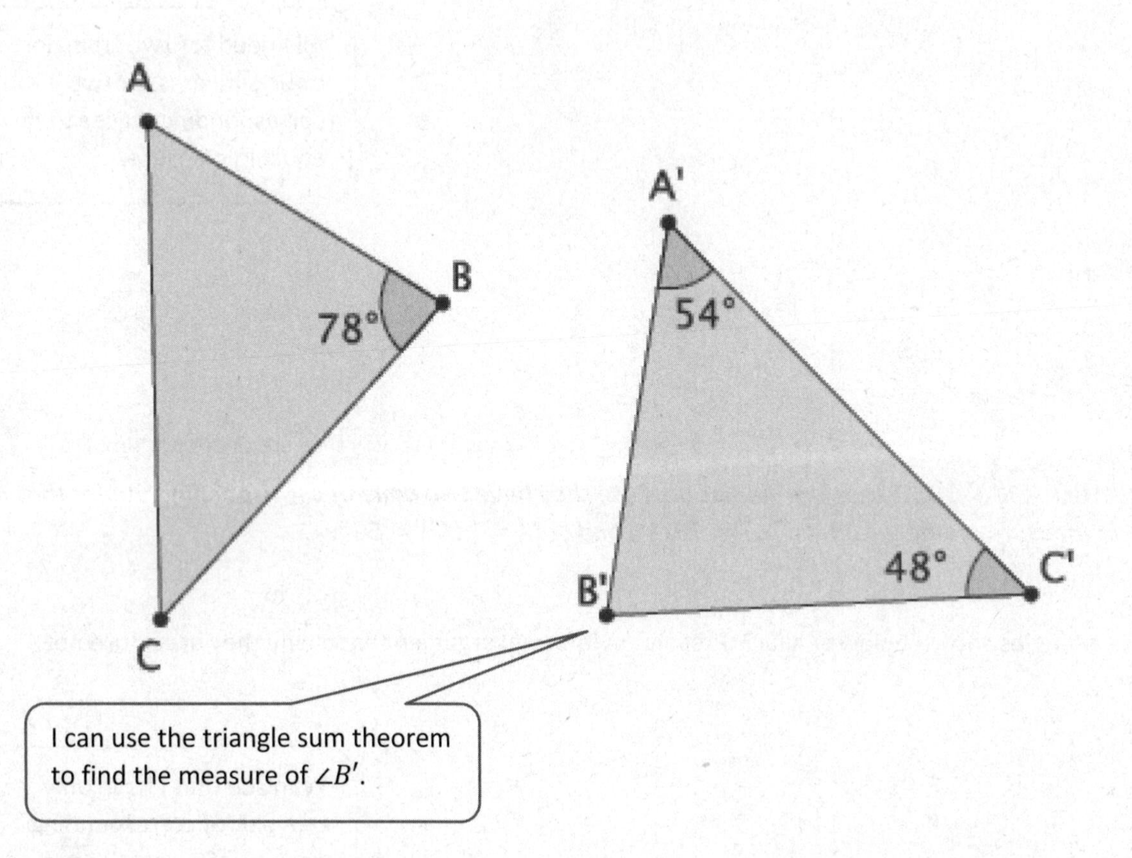

We do not know if $\triangle ABC$ is similar to $\triangle A'B'C'$. We can use the triangle sum theorem to find out th $|\angle B'| = 78°$, but we do not have any information about $|\angle A|$ or $|\angle C|$. To be considered similar, the tw triangles must have two pairs of corresponding angles that are equal in measure. In this problem, we only know the measures of one pair of corresponding angles.

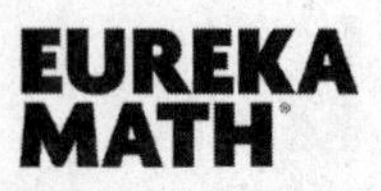

©2019 Great Minds®. eureka-math.org

1. Are the triangles shown below similar? Present an informal argument as to why they are or are not similar.

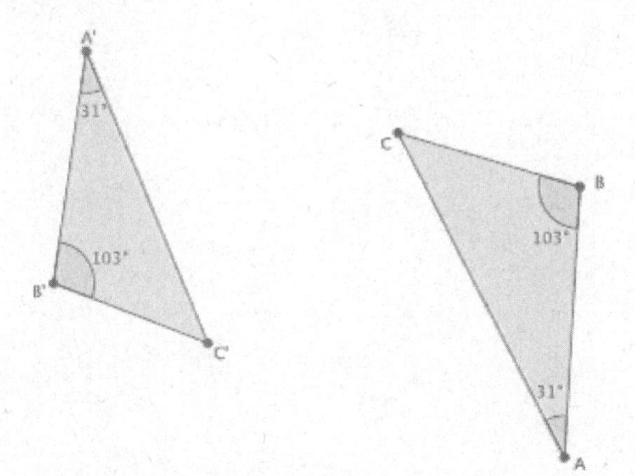

2. Are the triangles shown below similar? Present an informal argument as to why they are or are not similar.

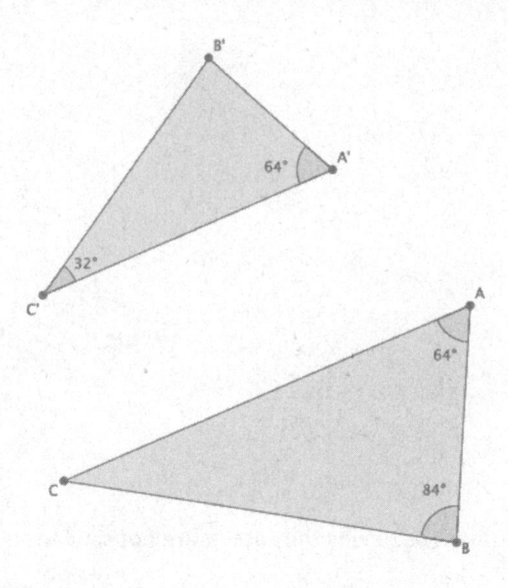

©2019 Great Minds®. eureka-math.org

3. Are the triangles shown below similar? Present an informal argument as to why they are or are not similar.

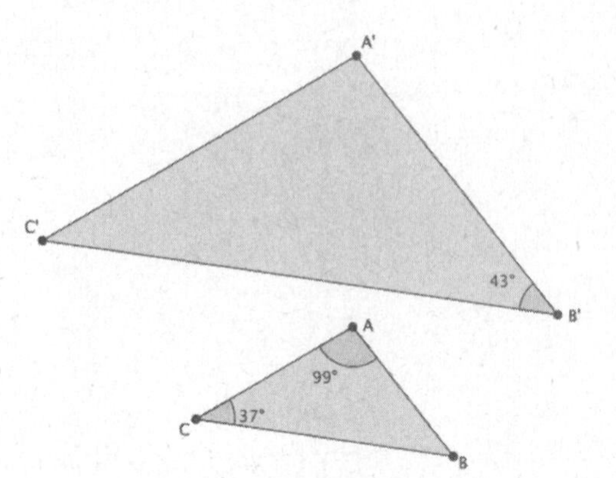

4. Are the triangles shown below similar? Present an informal argument as to why they are or are not similar.

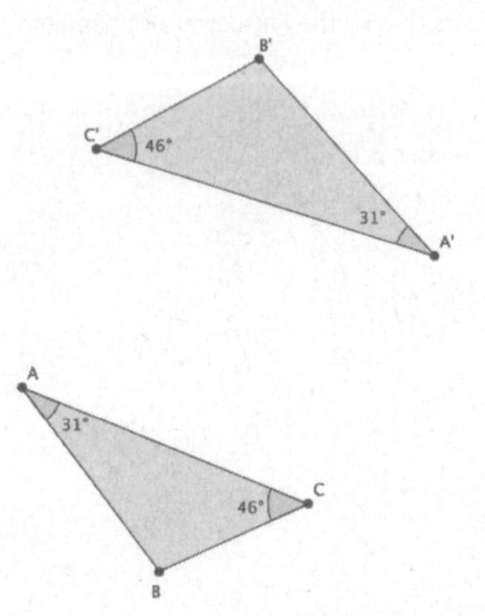

5. Are the triangles shown below similar? Present an informal argument as to why they are or are not similar.

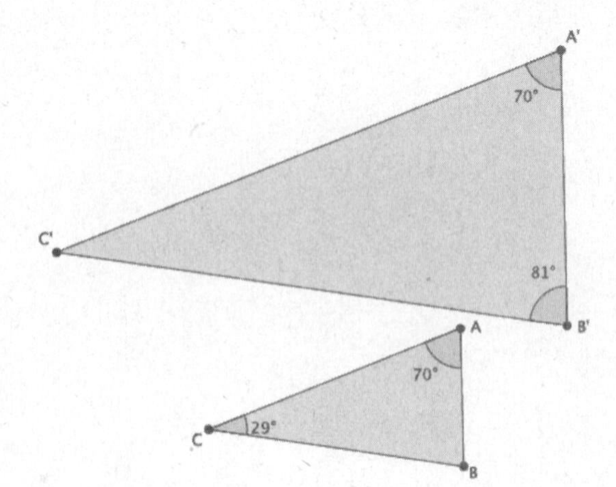

©2019 Great Minds®. eureka-math.org

6. Are the triangles shown below similar? Present an informal argument as to why they are or are not similar.

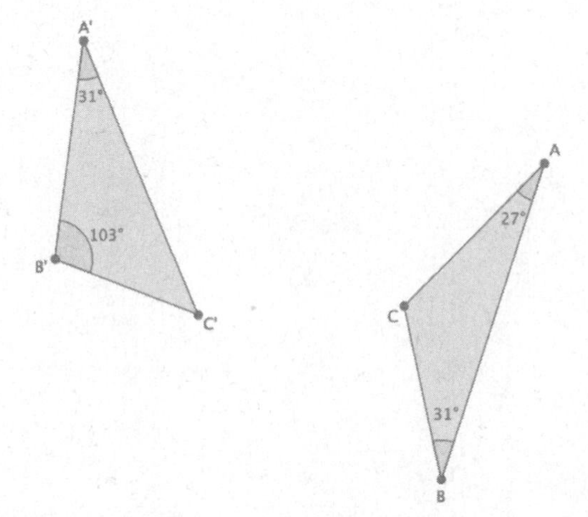

7. Are the triangles shown below similar? Present an informal argument as to why they are or are not similar.

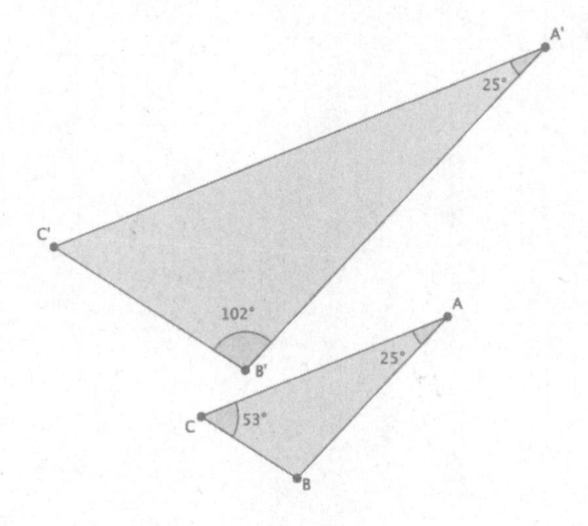

©2019 Great Minds®. eureka-math.org

Exercises

1. In the diagram below, you have $\triangle ABC$ and $\triangle AB'C'$. Use this information to answer parts (a)–(d).

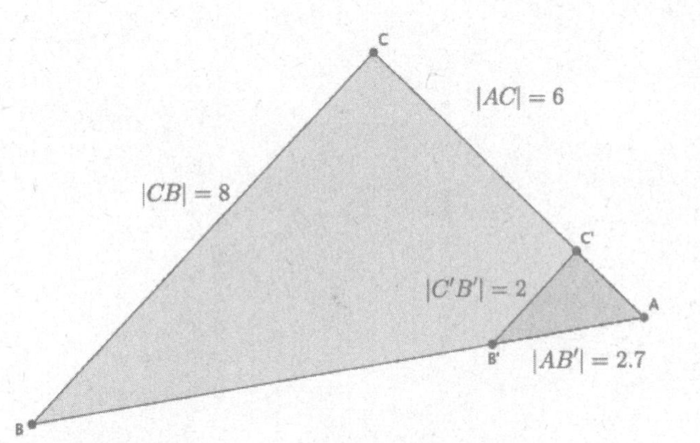

a. Based on the information given, is $\triangle ABC \sim \triangle AB'C'$? Explain.

b. Assume the line containing BC is parallel to the line containing $B'C'$. With this information, can you say that $\triangle ABC \sim \triangle AB'C'$? Explain.

c. Given that $\triangle ABC \sim \triangle AB'C'$, determine the length of side $\overline{AC'}$.

d. Given that $\triangle ABC \sim \triangle AB'C'$, determine the length of side $\overline{AB}$.

©2019 Great Minds®. eureka-math.org

2. In the diagram below, you have $\triangle ABC$ and $\triangle A'B'C'$. Use this information to answer parts (a)–(c).

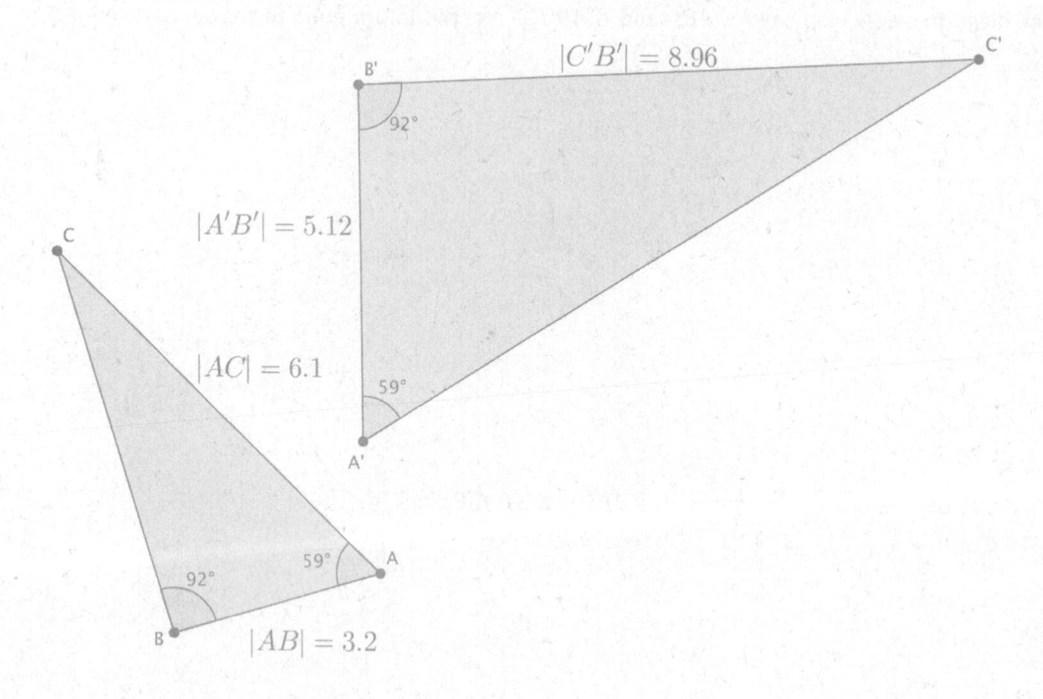

a. Based on the information given, is $\triangle ABC \sim \triangle A'B'C'$? Explain.

b. Given that $\triangle ABC \sim \triangle A'B'C'$, determine the length of side $\overline{A'C'}$.

c. Given that $\triangle ABC \sim \triangle A'B'C'$, determine the length of side $\overline{BC}$.

©2019 Great Minds®. eureka-math.org

3. In the diagram below, you have $\triangle ABC$ and $\triangle A'B'C'$. Use this information to answer the question below.

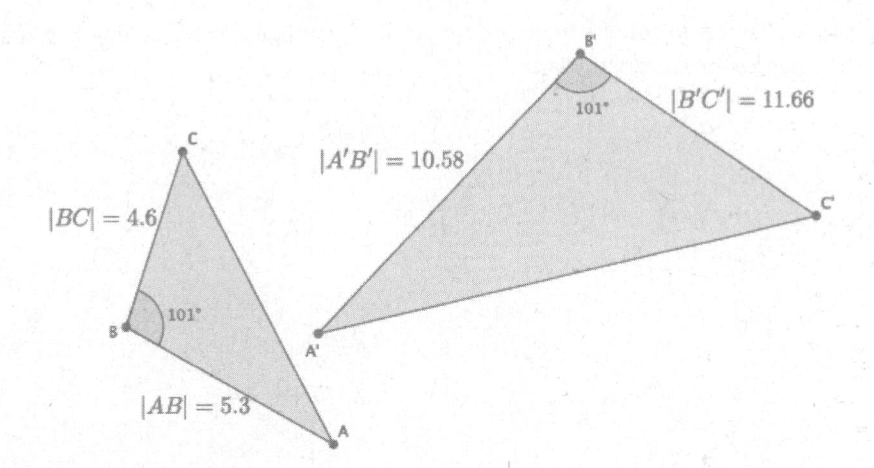

Based on the information given, is $\triangle ABC \sim \triangle A'B'C'$? Explain.

©2019 Great Minds®. eureka-math.org

Lesson Summary

Given just one pair of corresponding angles of a triangle as equal in measure, use the side lengths along the given angle to determine if the triangles are in fact similar.

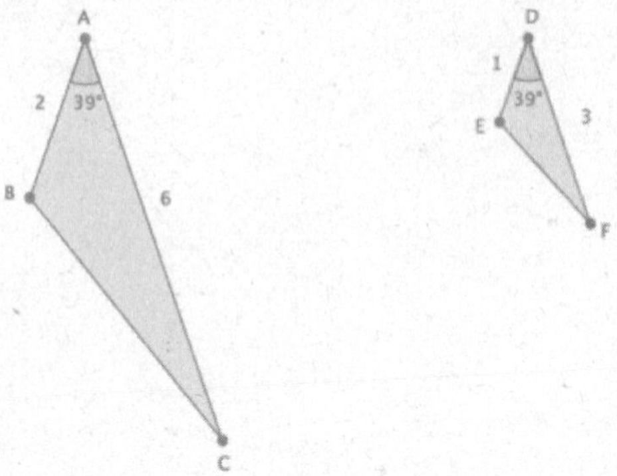

$|\angle A| = |\angle D|$ and $\frac{1}{2} = \frac{3}{6} = r$; therefore, $\triangle\ ABC \sim \triangle\ DEF$.

Given similar triangles, use the fact that ratios of corresponding sides are equal to find any missing measurements.

©2019 Great Minds®. eureka-math.org

Name ____ Date ________________

1. In the diagram below, you have $\triangle ABC$ and $\triangle A'B'C'$. Based on the information given, is $\triangle ABC \sim \triangle A'B'C'$? Explain.

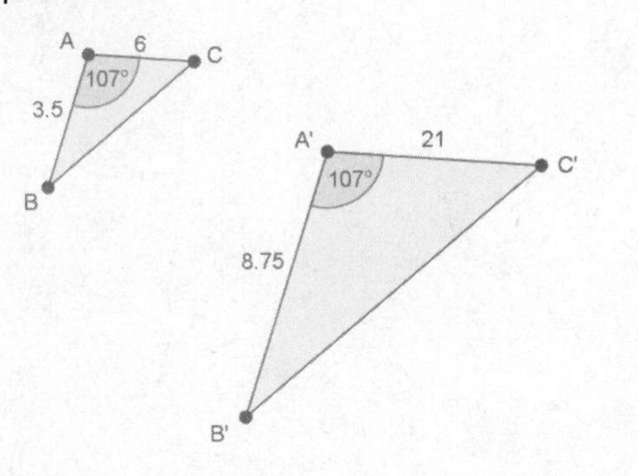

2. In the diagram below, $\triangle ABC \sim \triangle DEF$. Use the information to answer parts (a)–(b).

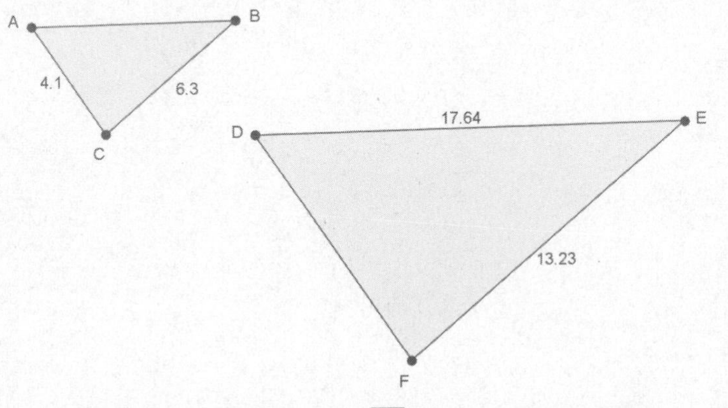

a. Determine the length of side $\overline{AB}$. Show work that leads to your answer.

b. Determine the length of side $\overline{DF}$. Show work that leads to your answer.

©2019 Great Minds®. eureka-math.org

1. In the diagram below, you have $\triangle ABC$ and $\triangle A'B'C'$. Use this information to answer parts (a)–(b).

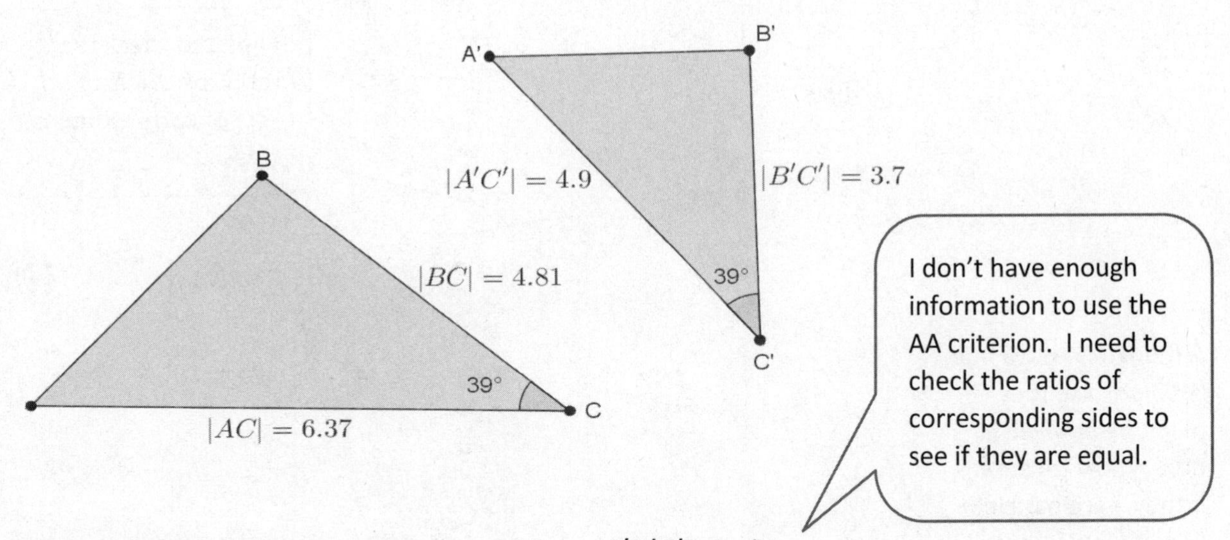

a. Based on the information given, is $\triangle ABC \sim \triangle A'B'C'$? Explain.

Yes, $\triangle ABC \sim \triangle A'B'C'$. Since there is only information about one pair of corresponding angles being equal in measure, then the corresponding sides must be checked to see if their ratios are equal.

$$\frac{4.81}{6.37} = \frac{3.7}{4.9}$$
$$0.755\ldots = 0.755\ldots$$

Since the values of these ratios are equal, approximately 0.755, the triangles are similar.

After I set up the ratio, I need to find the value of x that makes the fractions equivalent.

b. Assume the length of side $\overline{AB}$ is 4.03. What is the length of side $\overline{A'B'}$?

Let x represent the length of side $\overline{A'B'}$.

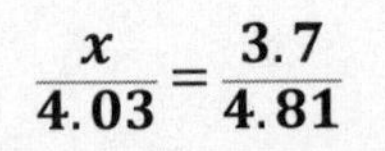

$$\frac{x}{4.03} = \frac{3.7}{4.81}$$

We are looking for the value of x that makes the fractions equivalent. Therefore, $4.81x = 14.911$, and $x = 3.1$. The length of side $\overline{A'B'}$ is 3.1.

©2019 Great Minds®. eureka-math.org

2. In the diagram below, you have $\triangle ABC$ and $\triangle A'BC'$. Based on the information given, is $\triangle ABC \sim \triangle A'BC'$? Explain.

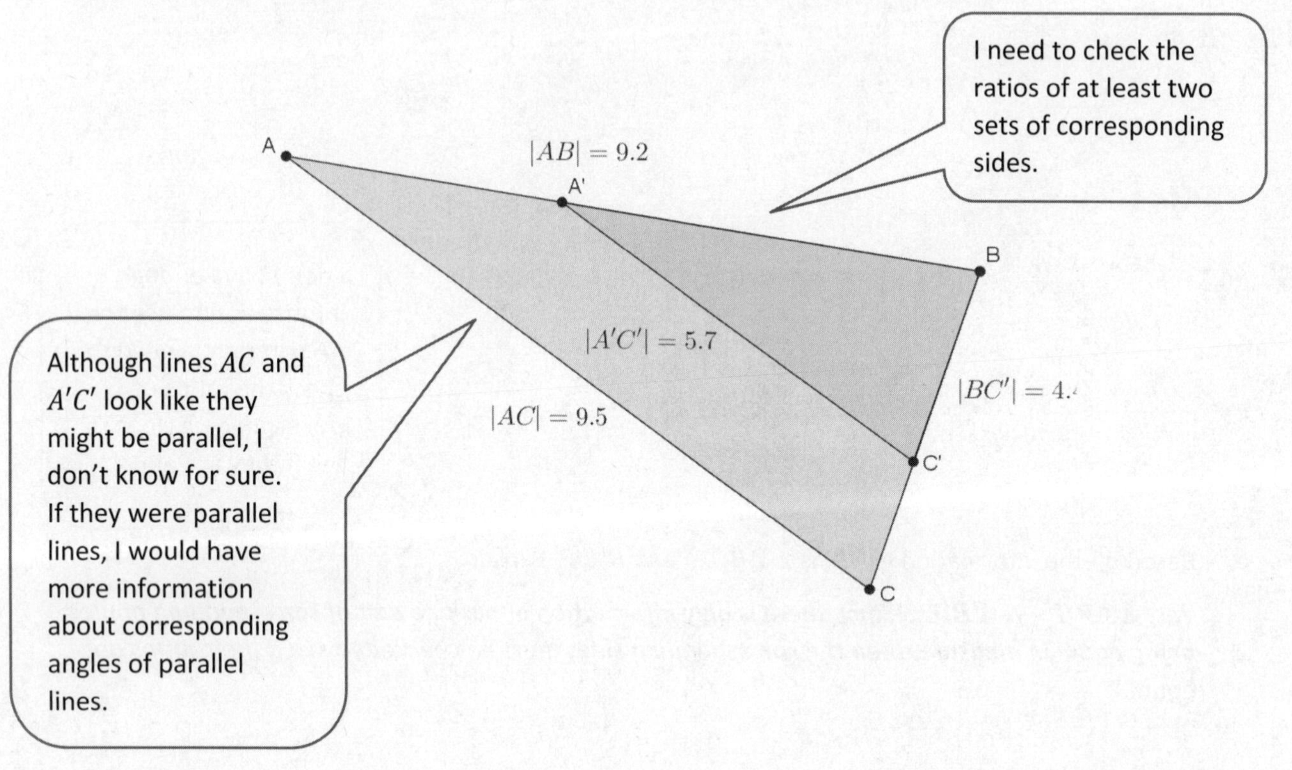

Since both triangles have a common vertex, then $|\angle B| = |\angle B|$. This means that the measure of $\angle B$ in $\triangle ABC$ is equal to the measure of $\angle B$ in $\triangle A'BC'$. However, there is not enough information provided to determine if the triangles are similar. We would need information about a pair of corresponding angles or more information about the side lengths of each of the triangles.

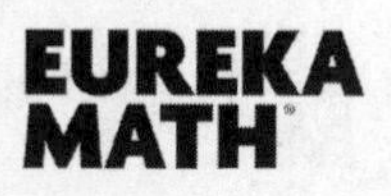

©2019 Great Minds®. eureka-math.org

1. In the diagram below, you have $\triangle ABC$ and $\triangle A'B'C'$. Use this information to answer parts (a)–(b).

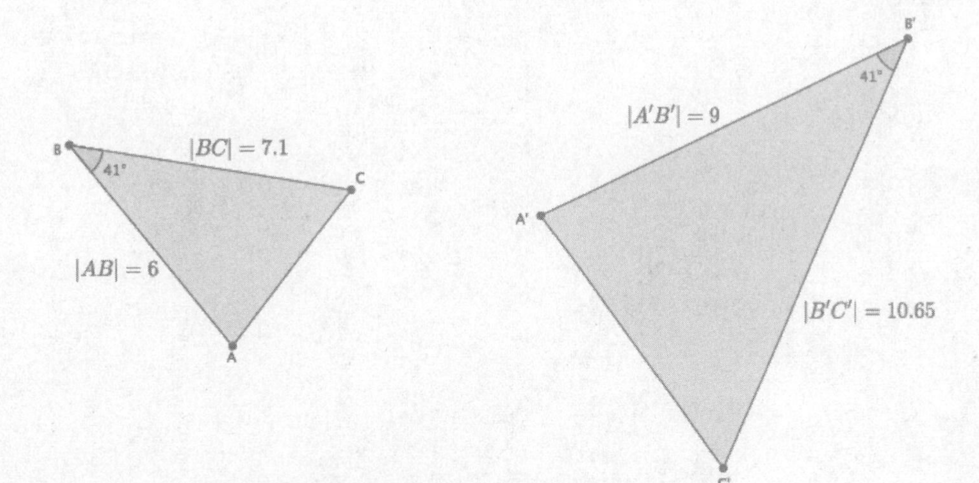

a. Based on the information given, is $\triangle ABC \sim \triangle A'B'C'$? Explain.

b. Assume the length of side $\overline{AC}$ is 4.3. What is the length of side $\overline{A'C'}$?

2. In the diagram below, you have $\triangle ABC$ and $\triangle AB'C'$. Use this information to answer parts (a)–(d).

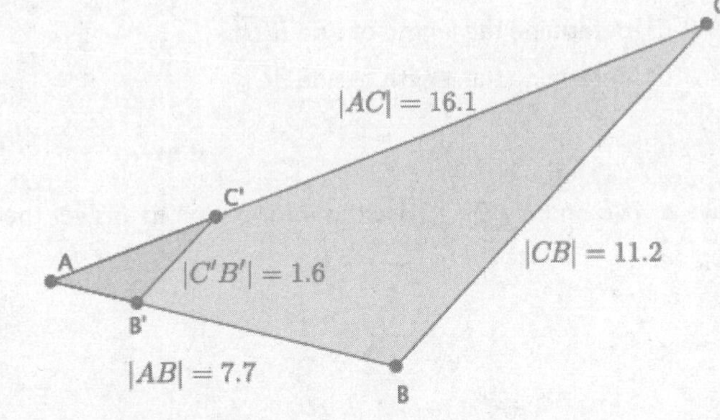

a. Based on the information given, is $\triangle ABC \sim \triangle AB'C'$? Explain.

b. Assume the line containing $\overline{BC}$ is parallel to the line containing $\overline{B'C'}$. With this information, can you say that $\triangle ABC \sim \triangle AB'C'$? Explain.

c. Given that $\triangle ABC \sim \triangle AB'C'$, determine the length of side $\overline{AC'}$.

d. Given that $\triangle ABC \sim \triangle AB'C'$, determine the length of side $\overline{AB'}$.

©2019 Great Minds®. eureka-math.org

3. In the diagram below, you have $\triangle ABC$ and $\triangle A'B'C'$. Use this information to answer parts (a)–(c).

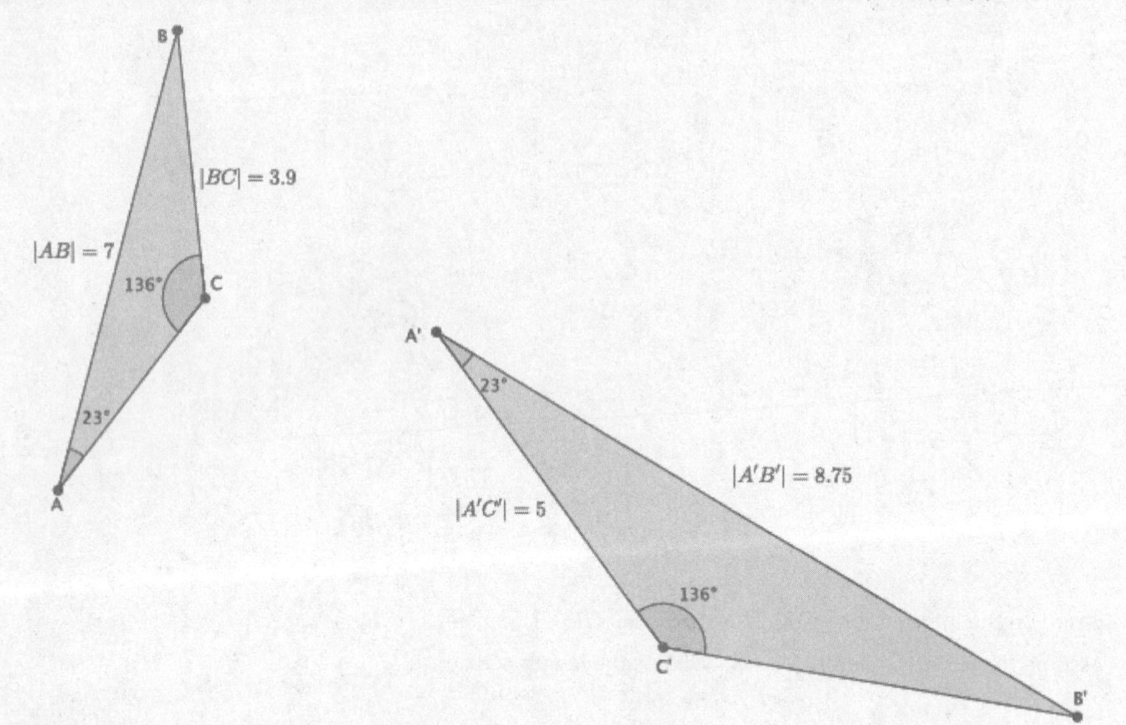

a. Based on the information given, is $\triangle ABC \sim \triangle A'B'C'$? Explain.

b. Given that $\triangle ABC \sim \triangle A'B'C'$, determine the length of side $\overline{B'C'}$.

c. Given that $\triangle ABC \sim \triangle A'B'C'$, determine the length of side $\overline{AC}$.

4. In the diagram below, you have $\triangle ABC$ and $\triangle AB'C'$. Use this information to answer the question below.

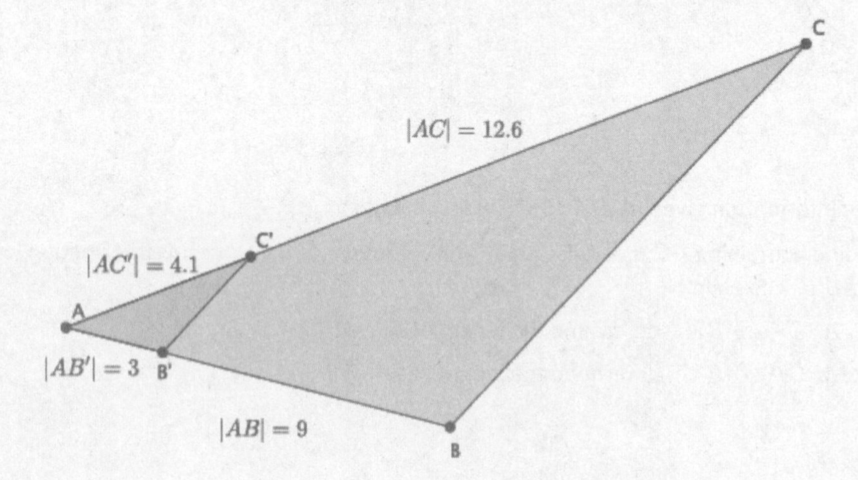

Based on the information given, is $\triangle ABC \sim \triangle AB'C'$? Explain.

EUREKA MATH
©2019 Great Minds®. eureka-math.org

5. In the diagram below, you have $\triangle ABC$ and $\triangle A'B'C'$. Use this information to answer parts (a)–(b).

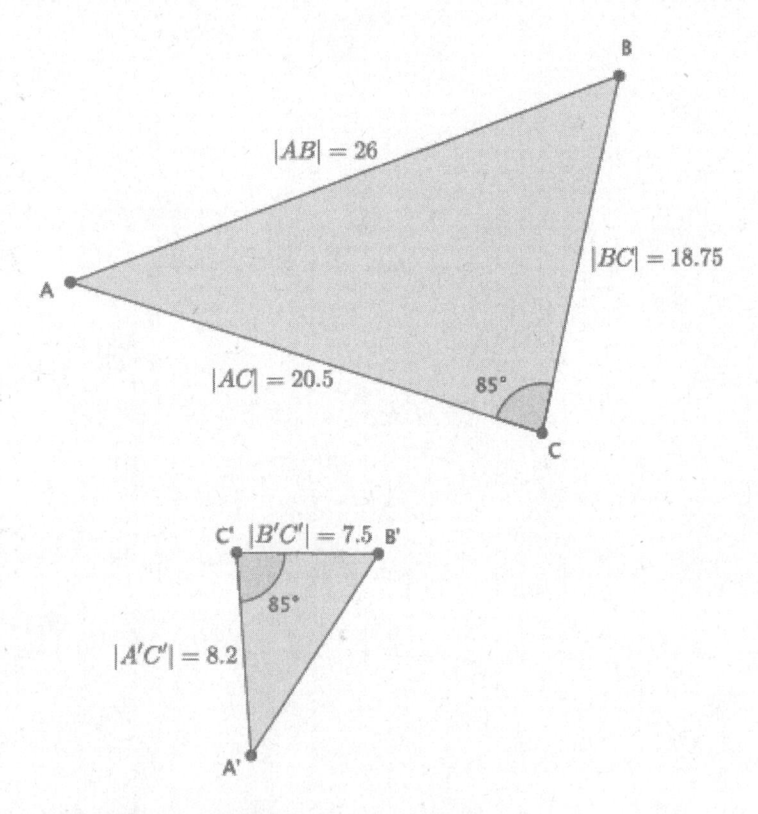

a. Based on the information given, is $\triangle ABC \sim \triangle A'B'C'$? Explain.

b. Given that $\triangle ABC \sim \triangle A'B'C'$, determine the length of side $\overline{A'B'}$.

©2019 Great Minds®. eureka-math.org

Example

Not all flagpoles are perfectly *upright* (i.e., perpendicular to the ground). Some are oblique (i.e., neither parallel nor at a right angle, slanted). Imagine an oblique flagpole in front of an abandoned building. The question is, can we use sunlight and shadows to determine the length of the flagpole?

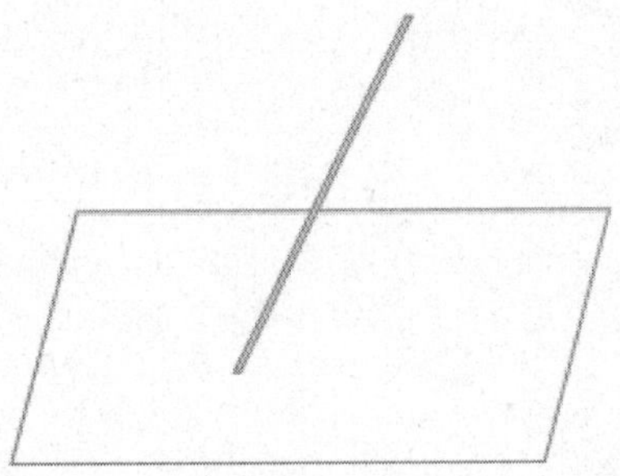

Assume that we know the following information: The length of the shadow of the flagpole is 15 feet. There is a mark on the flagpole 3 feet from its base. The length of the shadow of this three-foot portion of the flagpole is 1.7 feet.

©2019 Great Minds®. eureka-math.org

Mathematical Modeling Exercises 1–3

1. You want to determine the approximate height of one of the tallest buildings in the city. You are told that if you place a mirror some distance from yourself so that you can see the top of the building in the mirror, then you can indirectly measure the height using similar triangles. Let point O be the location of the mirror so that the person shown can see the top of the building.

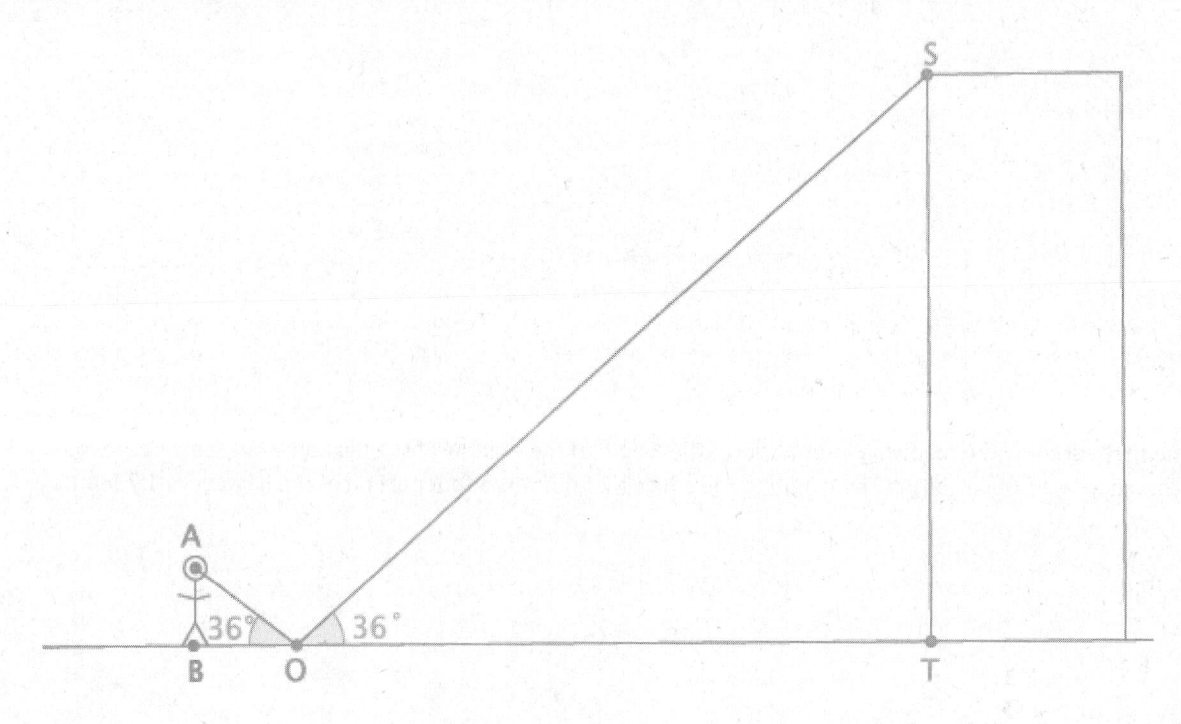

   a. Explain why $\triangle ABO \sim \triangle STO$.

   b. Label the diagram with the following information: The distance from eye level straight down to the ground is 5.3 feet. The distance from the person to the mirror is 7.2 feet. The distance from the person to the base of the building is 1,750 feet. The height of the building is represented by x.

   c. What is the distance from the mirror to the building?

©2019 Great Minds®. eureka-math.org

d. Do you have enough information to determine the approximate height of the building? If yes, determine the approximate height of the building. If not, what additional information is needed?

2. A geologist wants to determine the distance across the widest part of a nearby lake. The geologist marked off specific points around the lake so that the line containing $\overline{DE}$ would be parallel to the line containing $\overline{BC}$. The segment BC is selected specifically because it is the widest part of the lake. The segment DE is selected specifically because it is a short enough distance to easily measure. The geologist sketched the situation as shown below.

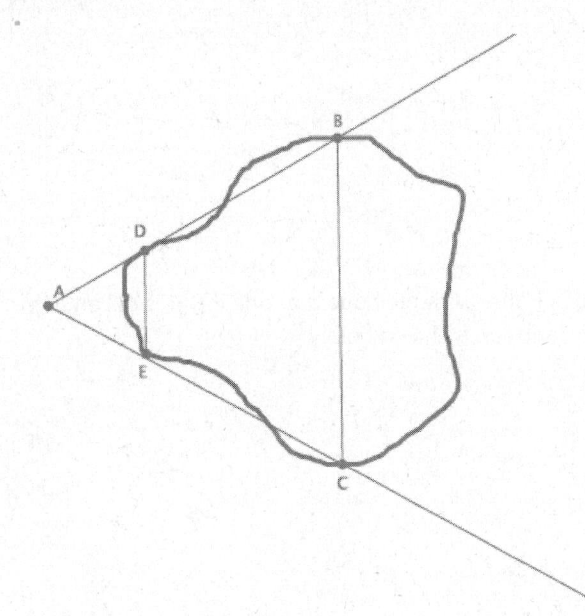

a. Has the geologist done enough work so far to use similar triangles to help measure the widest part of the lake? Explain.

©2019 Great Minds®. eureka-math.org

b. The geologist has made the following measurements: $|DE| = 5$ feet, $|AE| = 7$ feet, and $|EC| = 15$ feet. Does she have enough information to complete the task? If so, determine the length across the widest part of the lake. If not, state what additional information is needed.

c. Assume the geologist could only measure a maximum distance of 12 feet. Could she still find the distance across the widest part of the lake? What would need to be done differently?

3. A tree is planted in the backyard of a house with the hope that one day it is tall enough to provide shade to cool the house. A sketch of the house, tree, and sun is shown below.

a. What information is needed to determine how tall the tree must be to provide the desired shade?

©2019 Great Minds®. eureka-math.org

b. Assume that the sun casts a shadow 32 feet long from a point on top of the house to a point in front of the house. The distance from the end of the house's shadow to the base of the tree is 53 feet. If the house is 16 feet tall, how tall must the tree get to provide shade for the house?

c. Assume that the tree grows at a rate of 2.5 feet per year. If the tree is now 7 feet tall, about how many years will it take for the tree to reach the desired height?

©2019 Great Minds®. eureka-math.org

Name ______________________________ Date ______________

Henry thinks he can figure out how high his kite is while flying it in the park. First, he lets out 150 feet of string and ties the string to a rock on the ground. Then, he moves from the rock until the string touches the top of his head. He stands up straight, forming a right angle with the ground. He wants to find out the distance from the ground to his kite. He draws the following diagram to illustrate what he has done.

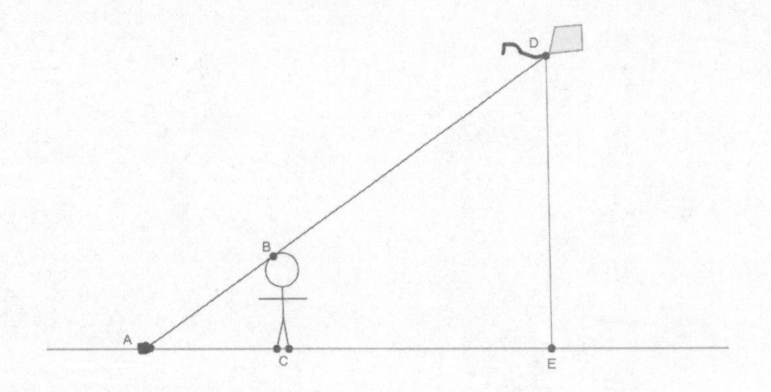

a. Has Henry done enough work so far to use similar triangles to help measure the height of the kite? Explain.

b. Henry knows he is $5\frac{1}{2}$ feet tall. Henry measures the string from the rock to his head and finds it to be 8 feet. Does he have enough information to determine the height of the kite? If so, find the height of the kite. If not, state what other information would be needed.

©2019 Great Minds®. eureka-math.org

1. There is a statue of your school's mascot in front of your school. You want to find out how tall the statue is, but it is too tall to measure directly. The following diagram represents the situation.

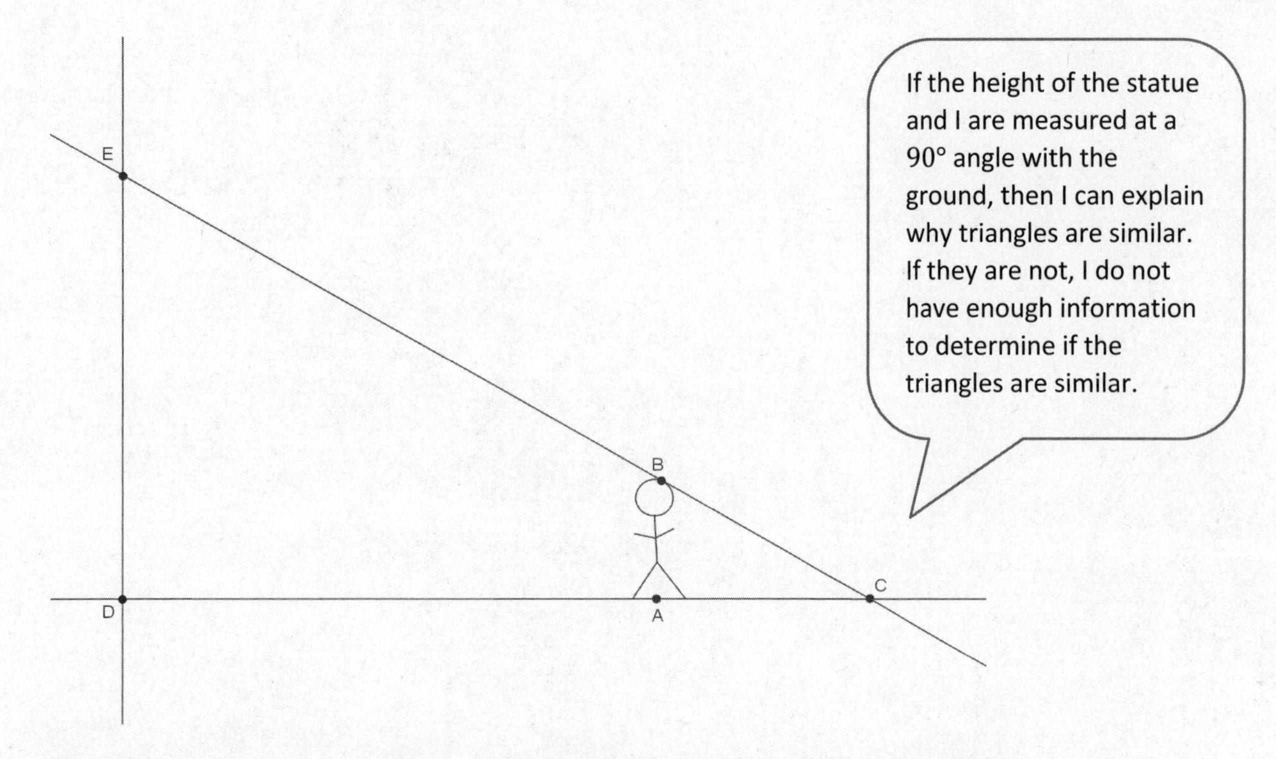

Describe the triangles in the situation, and explain how you know whether or not they are similar.

There are two triangles in the diagram, one formed by the statue and the shadow it casts, $\triangle EDC$, and another formed by the person and his shadow, $\triangle BAC$. The triangles are similar if the height of the statue is measured at a $90°$ angle with the ground and if the person standing forms a $90°$ angle with the ground. We know that $\angle ACB$ is an angle common to both triangles. If $|\angle EDC| = |\angle BAC| = 90°$, then $\triangle EDC \sim \triangle BAC$ by the AA criterion.

2. Assume $\triangle EDC \sim \triangle BAC$. If the statue casts a shadow 18 feet long and you are 5 feet tall and cast a shadow of 7 feet, find the height of the statue.

Let x represent the height of the statue; then

$$\frac{x}{5} = \frac{18}{7}.$$

Since I know the triangles are similar, I can set up a ratio of corresponding sides.

We are looking for the value of x that makes the fractions equivalent. Therefore, $7x = 90$, and $x = \frac{90}{7}$. The statue is about 13 feet tall.

©2019 Great Minds®. eureka-math.org

1. The world's tallest living tree is a redwood in California. It's about 370 feet tall. In a local park, there is a very tall tree. You want to find out if the tree in the local park is anywhere near the height of the famous redwood.

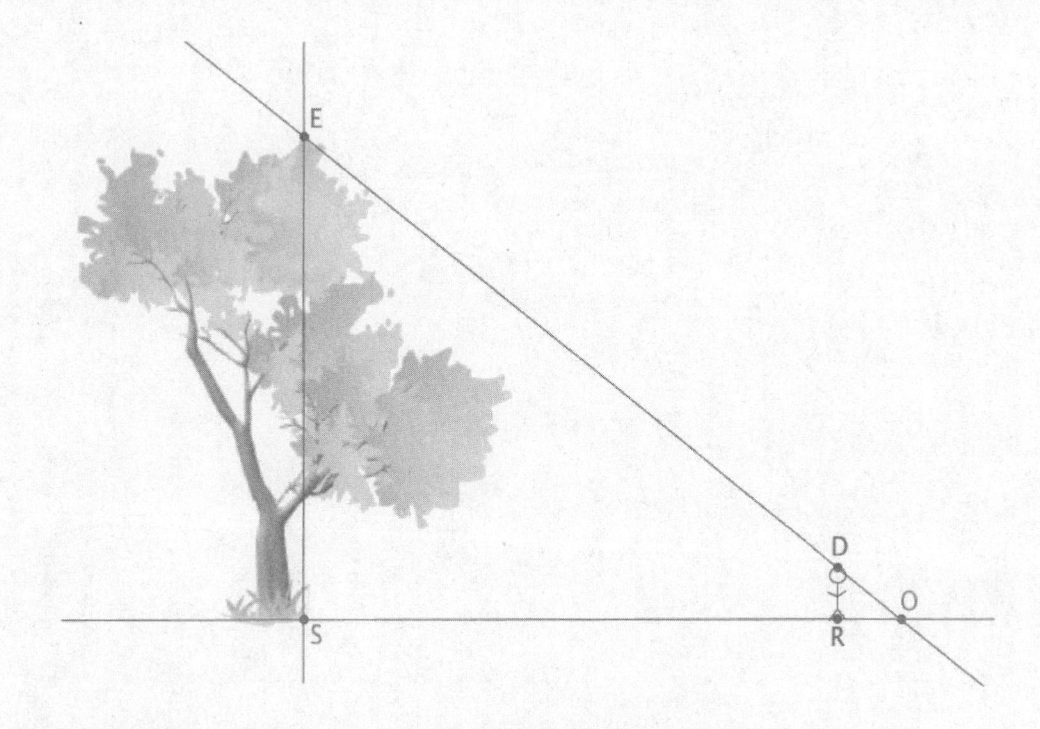

   a. Describe the triangles in the diagram, and explain how you know they are similar or not.

   b. Assume $\triangle ESO \sim \triangle DRO$. A friend stands in the shadow of the tree. He is exactly 5.5 feet tall and casts a shadow of 12 feet. Is there enough information to determine the height of the tree? If so, determine the height. If not, state what additional information is needed.

   c. Your friend stands exactly 477 feet from the base of the tree. Given this new information, determine about how many feet taller the world's tallest tree is compared to the one in the local park.

   d. Assume that your friend stands in the shadow of the world's tallest redwood, and the length of his shadow is just 8 feet long. How long is the shadow cast by the tree?

2. A reasonable skateboard ramp makes a 25° angle with the ground. A two-foot-tall ramp requires about 4.3 feet of wood along the base and about 4.7 feet of wood from the ground to the top of the two-foot height to make the ramp.

   a. Sketch a diagram to represent the situation.

   b. Your friend is a daredevil and has decided to build a ramp that is 5 feet tall. What length of wood is needed to make the base of the ramp? Explain your answer using properties of similar triangles.

   c. What length of wood is required to go from the ground to the top of the 5-foot height to make the ramp? Explain your answer using properties of similar triangles.

©2019 Great Minds®. eureka-math.org

Exercises

Use the Pythagorean theorem to determine the unknown length of the right triangle.

1. Determine the length of side c in each of the triangles below.

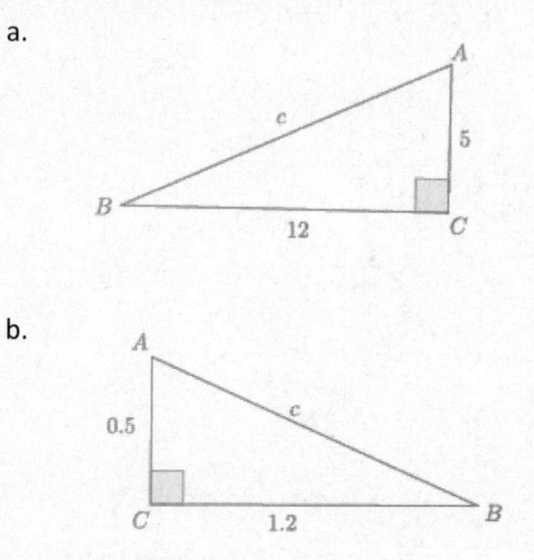

2. Determine the length of side b in each of the triangles below.

a.

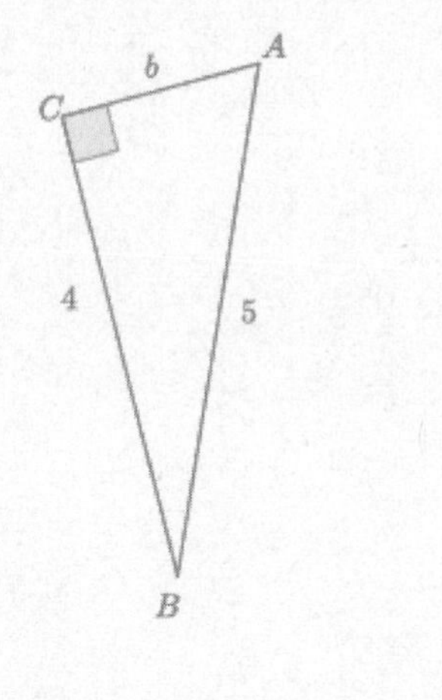

b.

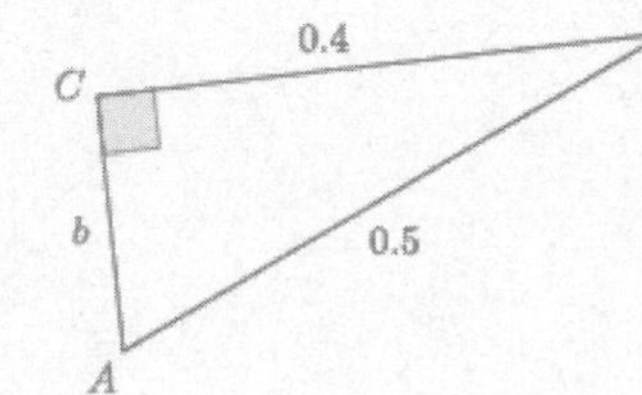

©2019 Great Minds®. eureka-math.org

3. Determine the length of $\overline{QS}$. (Hint: Use the Pythagorean theorem twice.)

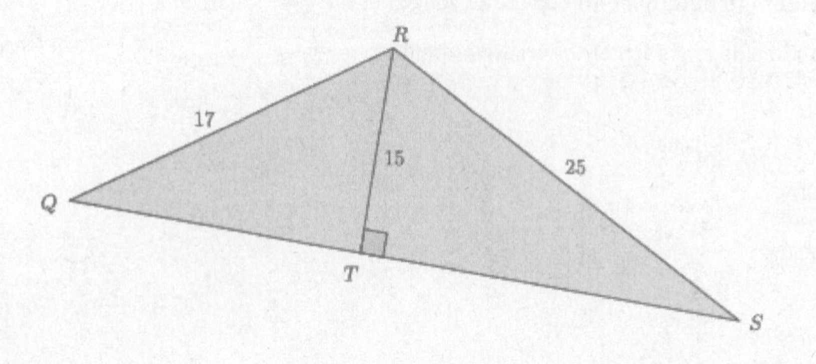

©2019 Great Minds®. eureka-math.org

Name ____________________ Date ____________

Determine the length of side $\overline{BD}$ in the triangle below.

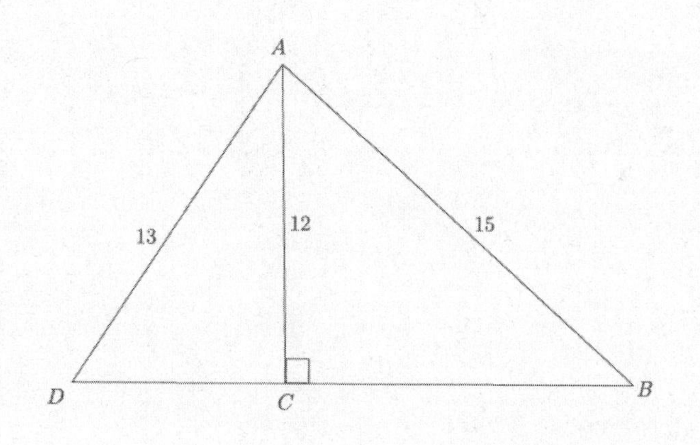

©2019 Great Minds®. eureka-math.org

Use the Pythagorean theorem to determine the unknown length of the right triangle.

1. Determine the length of side c in the triangle below.

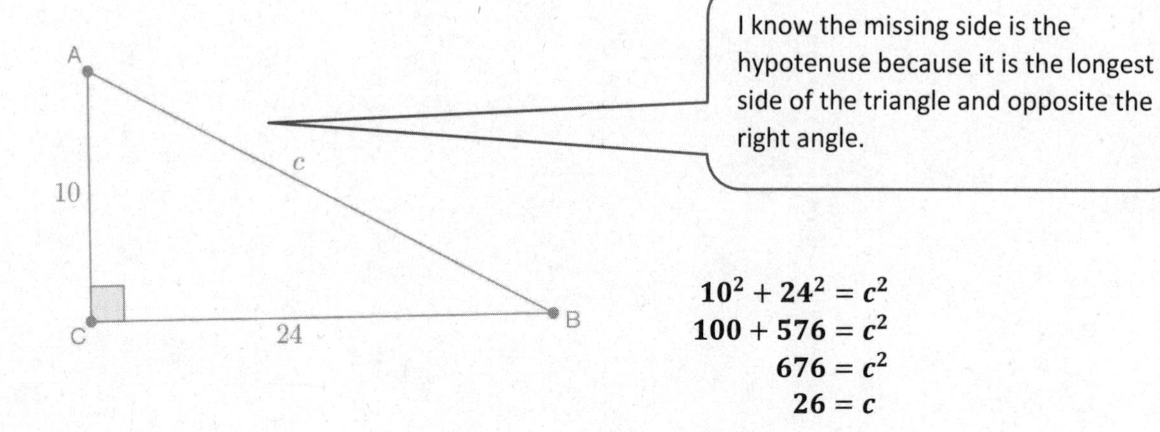

$$\begin{aligned} 10^2 + 24^2 &= c^2 \\ 100 + 576 &= c^2 \\ 676 &= c^2 \\ 26 &= c \end{aligned}$$

2. Determine the length of side b in the triangle below.

A
2.6
b
B
C
1

The side lengths are a tenth of the side lengths in problem one. I can multiply the side lengths by ten to make them whole numbers. That will make the problem easier.

$$\begin{aligned} 1^2 + b^2 &= 2.6^2 \\ 1 + b^2 &= 6.76 \\ 1 - 1 + b^2 &= 6.76 - 1 \\ b^2 &= 5.76 \\ b &= 2.4 \end{aligned}$$

©2019 Great Minds®. eureka-math.org

Use the Pythagorean theorem to determine the unknown length of the right triangle.

1. Determine the length of side c in each of the triangles below.

   a.

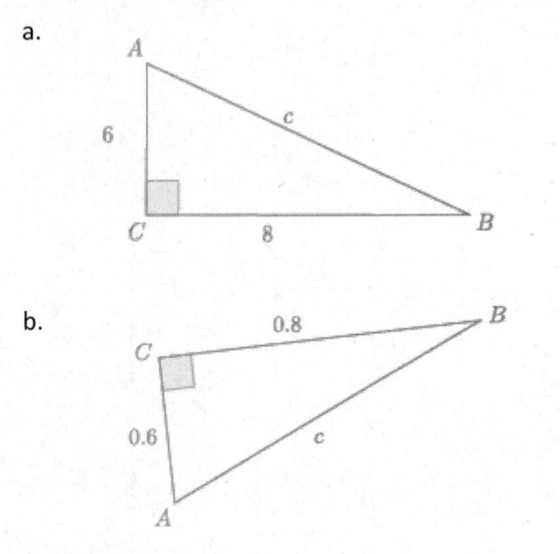

   b.

2. Determine the length of side a in each of the triangles below.

   a.

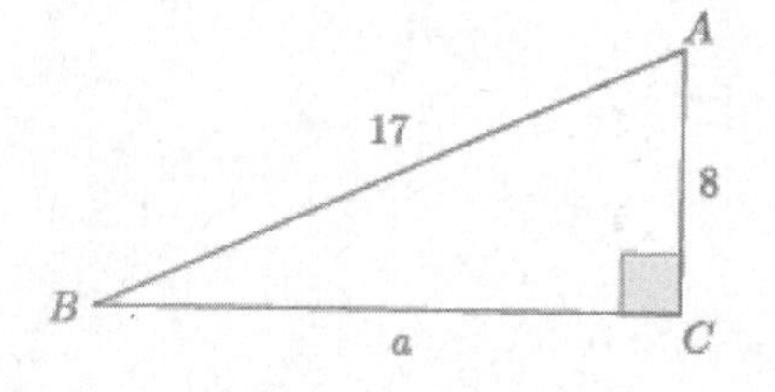

   b.

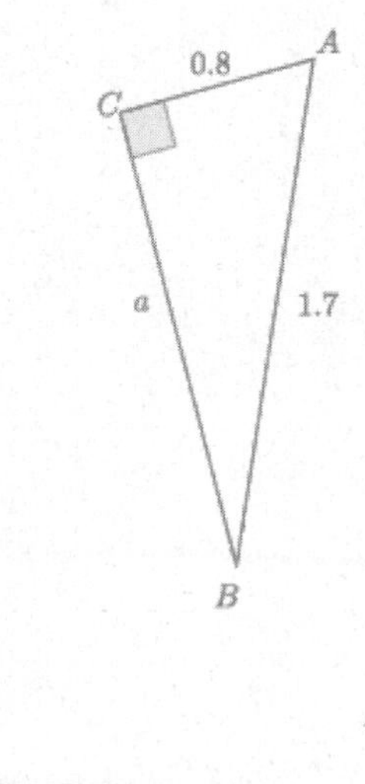

©2019 Great Minds®. eureka-math.org

3. Determine the length of side b in each of the triangles below.

a.

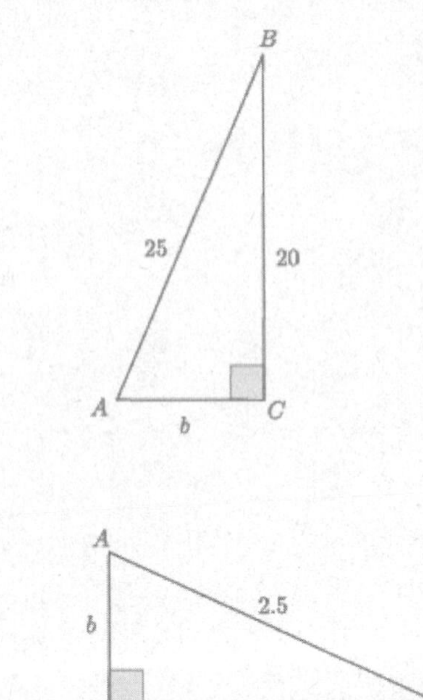

b.

A
2.5
b
C
2
B

4. Determine the length of side a in each of the triangles below.

a.

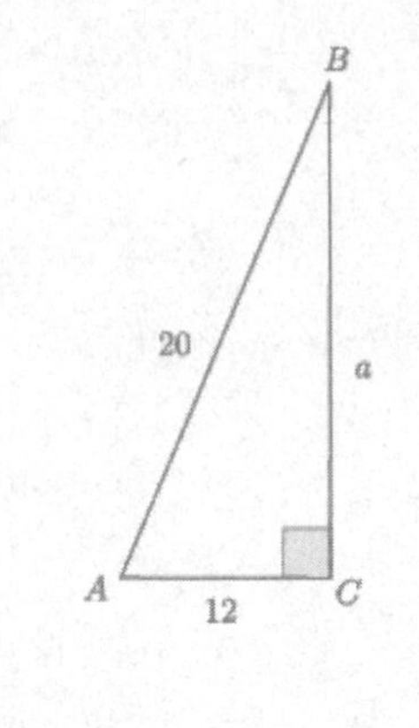

b.

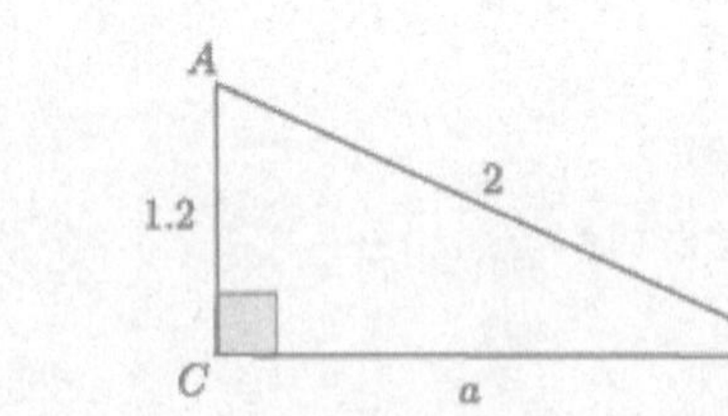

5. What did you notice in each of the pairs of Problems 1–4? How might what you noticed be helpful in solving problems like these?

©2019 Great Minds®. eureka-math.org

Exercises

1. The numbers in the diagram below indicate the units of length of each side of the triangle. Is the triangle shown below a right triangle? Show your work, and answer in a complete sentence.

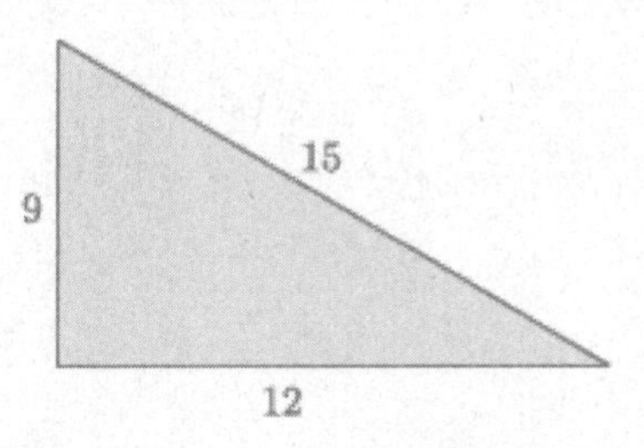

2. The numbers in the diagram below indicate the units of length of each side of the triangle. Is the triangle shown below a right triangle? Show your work, and answer in a complete sentence.

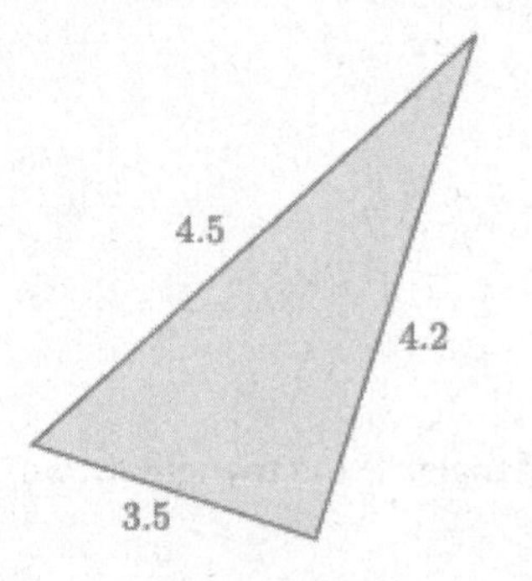

3. The numbers in the diagram below indicate the units of length of each side of the triangle. Is the triangle shown below a right triangle? Show your work, and answer in a complete sentence.

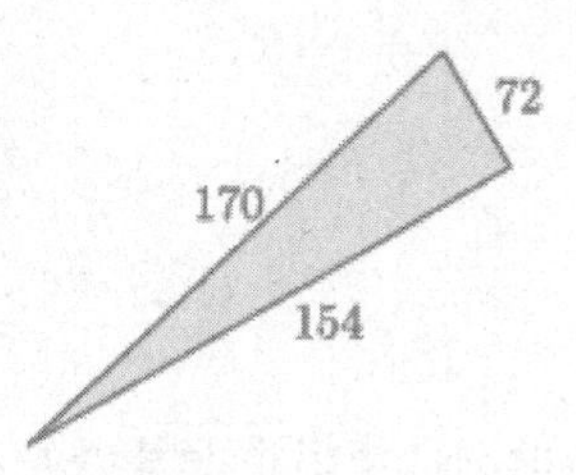

©2019 Great Minds®. eureka-math.org

4. The numbers in the diagram below indicate the units of length of each side of the triangle. Is the triangle shown below a right triangle? Show your work, and answer in a complete sentence.

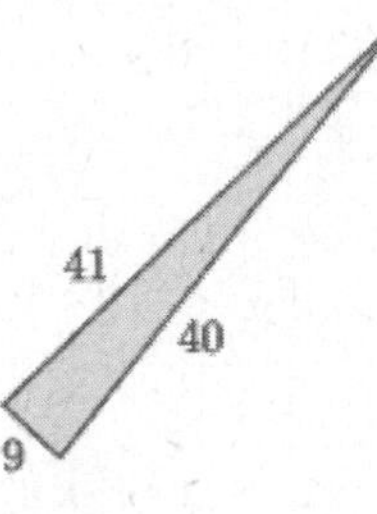

5. The numbers in the diagram below indicate the units of length of each side of the triangle. Is the triangle shown below a right triangle? Show your work, and answer in a complete sentence.

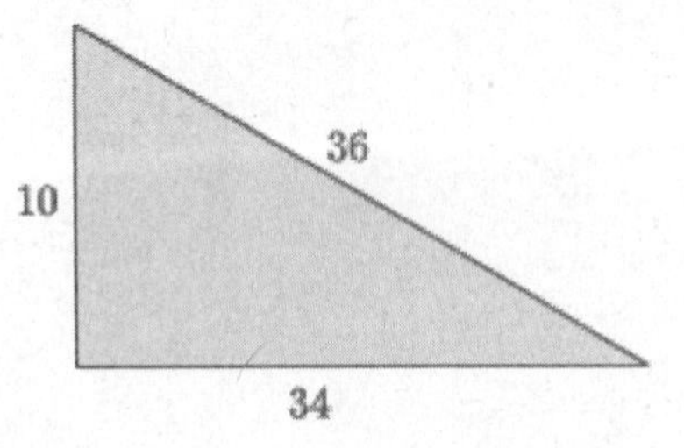

6. The numbers in the diagram below indicate the units of length of each side of the triangle. Is the triangle shown below a right triangle? Show your work, and answer in a complete sentence.

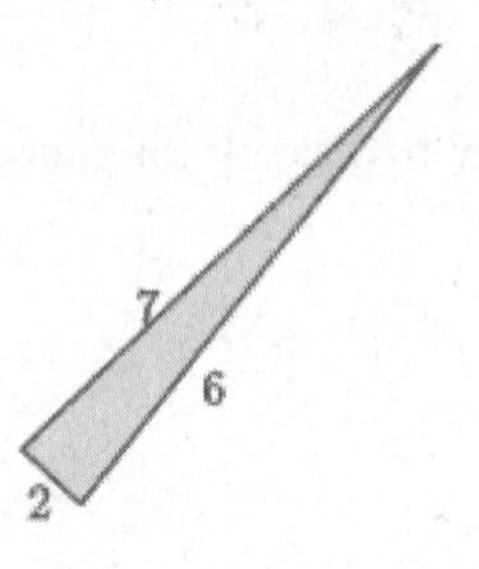

7. The numbers in the diagram below indicate the units of length of each side of the triangle. Is the triangle shown below a right triangle? Show your work, and answer in a complete sentence.

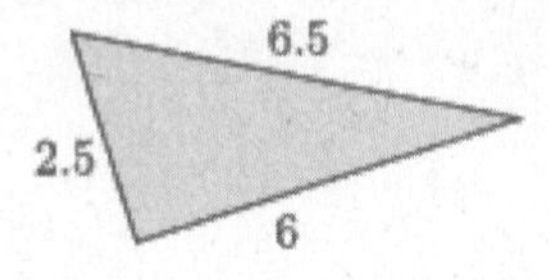

©2019 Great Minds®. eureka-math.org

Lesson Summary

The converse of the Pythagorean theorem states that if the side lengths of a triangle, a, b, c, satisfy $a^2 + b^2 = c^2$, then the triangle is a right triangle.

If the side lengths of a triangle, a, b, c, do not satisfy $a^2 + b^2 = c^2$, then the triangle is not a right triangle.

©2019 Great Minds®. eureka-math.org

Name ______________________________ Date ______________

1. The numbers in the diagram below indicate the lengths of the sides of the triangle. Bernadette drew the following triangle and claims it is a right triangle. How can she be sure?

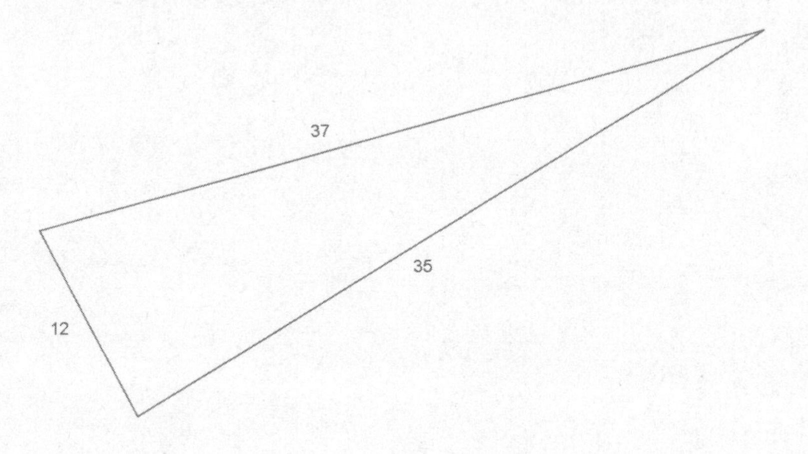

2. Do the lengths 5, 9, and 14 form a right triangle? Explain.

©2019 Great Minds®. eureka-math.org

1. The numbers in the diagram below indicate the units of length of each side of the triangle. Is the triangle shown below a right triangle? Show your work, and answer in a complete sentence.

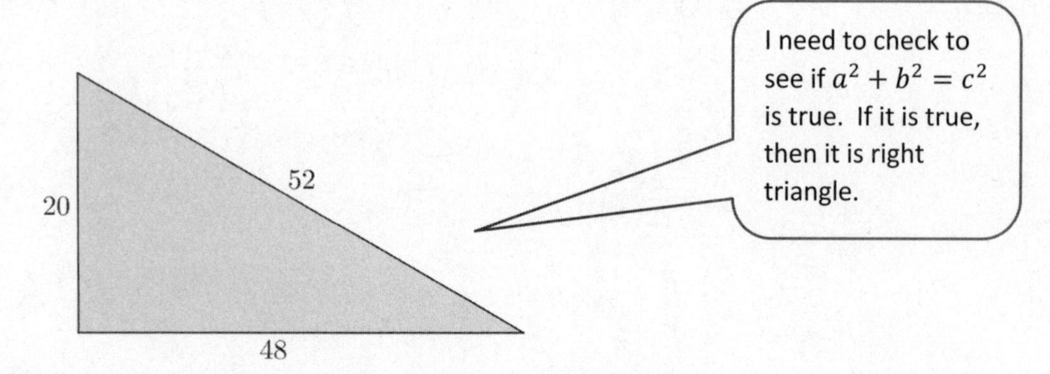

We need to check if $20^2 + 48^2 = 52^2$ is a true statement. The left side of the equation is equal to $2,704$. The right side of the equation is equal to $2,704$. That means $20^2 + 48^2 = 52^2$ is true, and the triangle shown is a right triangle.

2. The numbers in the diagram below indicate the units of length of each side of the triangle. Is the triangle shown below a right triangle? Show your work, and answer in a complete sentence.

We need to check if $24^2 + 30^2 = 40^2$ is a true statement. The left side of the equation is equal to $1,476$. The right side of the equation is equal to $1,600$. That means $24^2 + 30^2 = 40^2$ is not true, and the triangle shown is not a right triangle.

©2019 Great Minds®. eureka-math.org

1. The numbers in the diagram below indicate the units of length of each side of the triangle. Is the triangle shown below a right triangle? Show your work, and answer in a complete sentence.

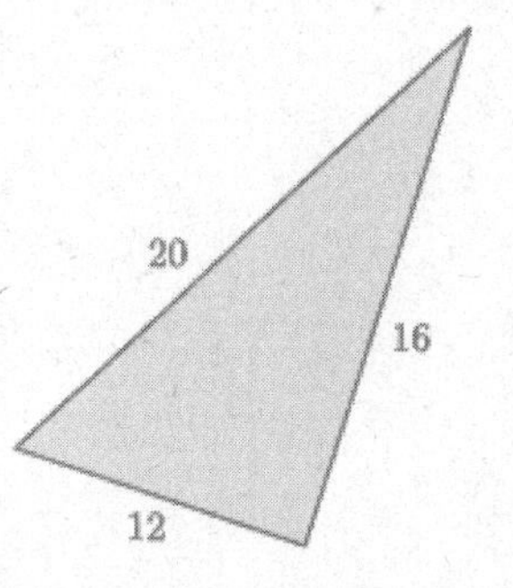

2. The numbers in the diagram below indicate the units of length of each side of the triangle. Is the triangle shown below a right triangle? Show your work, and answer in a complete sentence.

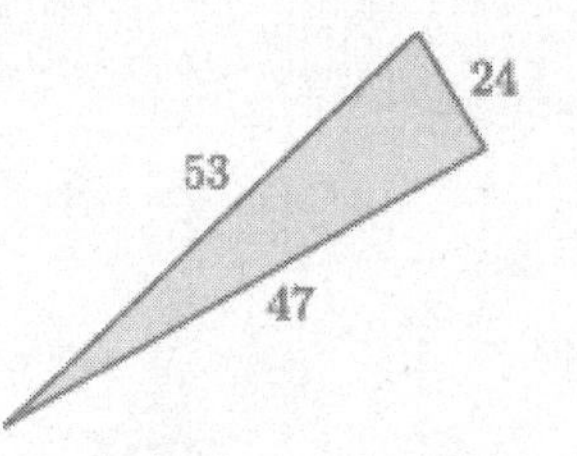

3. The numbers in the diagram below indicate the units of length of each side of the triangle. Is the triangle shown below a right triangle? Show your work, and answer in a complete sentence.

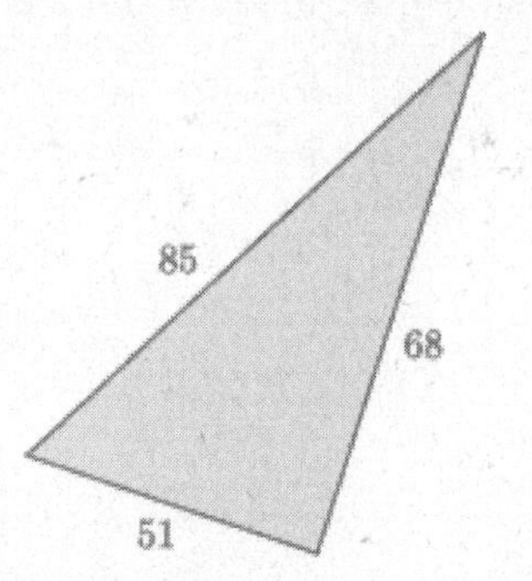

4. The numbers in the diagram below indicate the units of length of each side of the triangle. Sam said that the following triangle is a right triangle because $9 + 32 = 40$. Explain to Sam what he did wrong to reach this conclusion and what the correct solution is.

©2019 Great Minds®. eureka-math.org

5. The numbers in the diagram below indicate the units of length of each side of the triangle. Is the triangle shown below a right triangle? Show your work, and answer in a complete sentence.

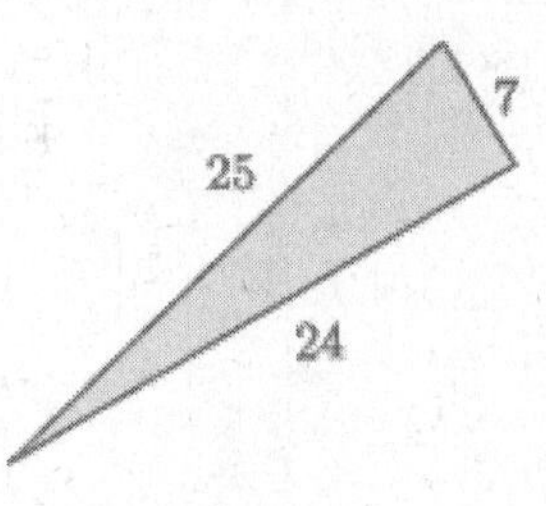

6. Jocelyn said that the triangle below is not a right triangle. Her work is shown below. Explain what she did wrong, and show Jocelyn the correct solution.

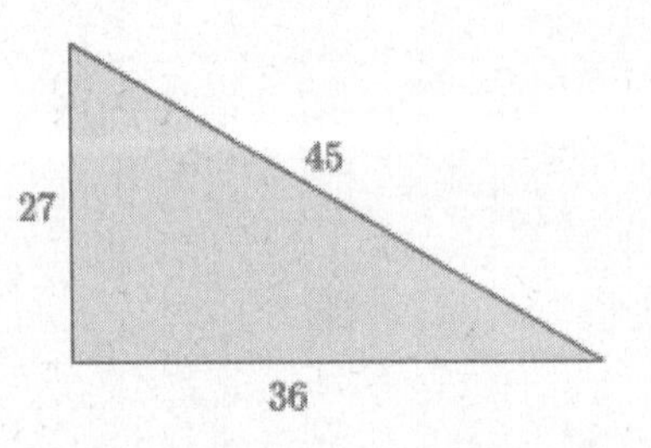

We need to check if $27^2 + 45^2 = 36^2$ is a true statement. The left side of the equation is equal to 2,754. The right side of the equation is equal to 1,296. That means $27^2 + 45^2 = 36^2$ is not true, and the triangle shown is not a right triangle.

©2019 Great Minds®. eureka-math.org

Credits

Great Minds® has made every effort to obtain permission for the reprinting of all copyrighted material. If any owner of copyrighted material is not acknowledged herein, please contact Great Minds for proper acknowledgment in all future editions and reprints of this module.

©2019 Great Minds®. eureka-math.org